D1382789

Laser Raman Spectroscopy

LASER RAMAN SPECTROSCOPY

A survey of interest primarily to chemists, and containing a comprehensive discussion of experiments on crystals

T. R. GILSON
Research Fellow, Department of Chemistry University of Southampton

P. J. HENDRA
Lecturer in Physical Chemistry, Department of Chemistry, University of Southampton

WILEY-INTERSCIENCE
A Division of John Wiley & Sons Ltd
LONDON NEW YORK SYDNEY TORONTO

Library of Congress Catalog Card No. 70-116651

ISBN 0 471 30200 7

Printed in Great Britain by
Dawson & Goodall Ltd.,
The Mendip Press, Bath

Contents

1. Introduction **1**
1.1 Electromagnetic Radiation 3
1.2 Interaction with Matter 8

2. Experimental **21**
2.1 Introduction 21
2.2 Basic Units of a Modern Raman Spectrometer . . . 24
2.3 Historical Development of Laser Raman Spectrometers . 35
2.4 Commercially Available Spectrometers 37
2.5 Performance of Spectrometers 50

3. Raman Intensities and Depolarization Ratios **61**
3.1 Treatment for Fluids 62
3.2 Treatment for Powders 78
3.3 Interpretation of Raman Intensity and Depolarization Measurements 79

4. Single-crystal Raman Spectroscopy **84**
4.1 Introduction 84
4.2 Vibrations and Symmetry 85
4.3 Elementary Practical Considerations 131
4.4 Determination of Sign of R_{ij} 138
4.5 Work in the Literature 139
4.6 Bibliography for Chapter 4 140

5. Raman Spectra of Gases **142**
5.1 Experimental 143
5.2 Results 146
5.3 'New' Experiments 147
5.4 Conclusion 152

6. Powders, Liquids, and Solutions **153**

7. Polymers **166**
7.1 Spectra recorded before the Advent of the Laser . . 166
7.2 Experimental Techniques 168
7.3 Raman Spectra of Specific Polymers 170
7.4 Conclusion 182

8. Miscellaneous Raman Experiments **183**
8.1 Raman Spectra of Adsorbed Layers 183
8.2 The Electronic Raman Effect 189
8.3 Non-linear Effects 191

9. Sources of Information **198**

Appendix 1: The Laser **201**

Appendix 2: Results from Group Theory **207**
A2.1 Tensor and Vector Activity of Translations and Rotations 207
A2.2 Character Tables and Raman Tensors 208
A2.3 Correlation Tables 234
A2.4 Possible Site Symmetries 247

References **254**

Index **262**

Preface

This book attempts to provide a logical, qualitative description of Raman scattering from principles which should be familiar to graduate chemists. An exception to this is the assumption of extended familiarity with infrared spectroscopy and molecular vibrations, both of which subjects are already adequately covered in the literature. The vibrational properties of single crystals are less adequately explained in introductory texts, and therefore their treatment is correspondingly extended here. The book should extend the knowledge of users of infrared spectroscopy, so that full advantage can be taken of the increasing applicability of Raman spectroscopy.

In the treatment of Raman intensities (Chapter 3) and molecular crystals (Chapter 4), the traditional order of presentation has been abandoned. This is because the authors believe that the theoretical background to these subjects must be appreciated before the simpler treatments can be safely used.

It is not intended that the book should be considered for undergraduate or postgraduate *course* teaching as such, but it will be of value to persons with a special interest in spectroscopy. In certain sections, an effort has been made to interest the industrial and analytical chemist. There is an extensive list of reviews and other sources of information in this and closely related fields, since the book is not intended to be a comprehensive review of the original literature.

The authors wish to express their gratitude to Professor I. R. Beattie who provided the facilities for much of the experimental work described here, and also for his continuing help, advice, and encouragement. Their thanks are also due to their colleagues in the Department of Chemistry for providing results prior to publication, and particularly to Mr. Malcolm Gall and Mr. Jack Loader for assistance in preparing Chapters 3 and 8 respectively. Any errors are, of course, the responsibility of the authors.

T.R.G.

Southampton, June 1969 P.J.H.

Foreword

Many of the physical methods used in chemistry determine separations between energy levels of one kind or another. Indeed, the average chemist sometimes finds it difficult to believe that matter can exhibit so many physical phenomena, all apparently (in the words of manufacturers' literature) of 'immense potential' and 'invaluable to every chemist'. The thought must sometimes arise: "Is it merely good for trade?".

In these circumstances it is pleasant to introduce not another unheard of set of energy levels but simply a slightly unusual way of determining the positions of a very familiar set—the vibrational and rotational levels of molecules. Many methods of investigating energy levels (e.g. ultraviolet, visible, and infrared absorption and emission, nuclear magnetic resonance, electron spin resonance, fluorescence, X-ray fluorescence, Mössbauer spectroscopy, etc.) are essentially resonance methods, and rely on the detection of particles or frequencies in some way *directly* indicative of a molecular or nuclear process. Raman spectroscopy, in common with some other techniques, is an inelastic scattering phenomenon, i.e. a small change is made to the energy and/or frequency of a quantum incident upon a sample, this change being related to the energy levels under examination. Since it is not a resonant process it is normally only very weakly allowed.

The Raman process is a *light* scattering phenomenon, first predicted from theoretical considerations by A. Smekal [*Naturwiss.*, **11,** 873 (1923)]. In 1928 C. V. Raman [*Nature*, **121,** 619 (1928)] examined the light scattered by dense media, and discovered the effect later named after him. The process may be described as follows. If a sample is irradiated with monochromatic radiation from, for example, a high-powered mercury discharge lamp, it is found that the incident radiation is weakly scattered. This scattering is caused by dust particles or inhomogeneities in the medium (Tyndall scattering), and to a weaker extent by the polarizability of the medium (Rayleigh scattering). Both processes cause the radiation to be scattered at the *same frequency* as that of the source. In addition to these processes, a third form of scattering occurs at frequencies different from that of the source, and this phenomenon is the subject of the present book. In the case of liquid chlorine irradiated with mercury light at 5461 Å, analysis of the scattered light with a spectrograph would show three 'lines' at 5300, 5461, and 5632 Å. The intensity ratios are roughly 1 : 1000 : 2, i.e. the radiation to the red side of the source line is stronger than that to the blue side. Conversion of these wavelengths to frequencies reveals that the two weak Raman lines are symmetrically placed about the source line at +556 and −556 cm^{-1} respectively, i.e. at the vibration frequency of the chlorine molecule. Turning to a molecule with a larger number of

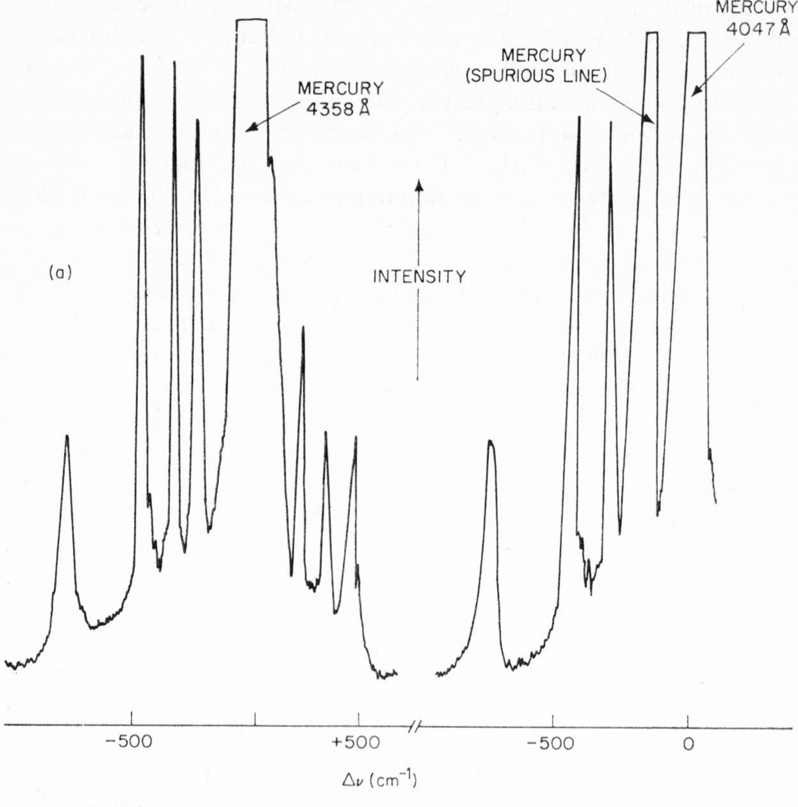

Figure F.1 Raman spectra recorded with a number of different source lines. All the graphs are microphotometer tracings from photographic plates

(a) Spectra from two mercury lines scattering simultaneously. A mercury line that is too weak to excite spectra but still very strong is also shown. Note that Stokes lines are stronger than anti-Stokes

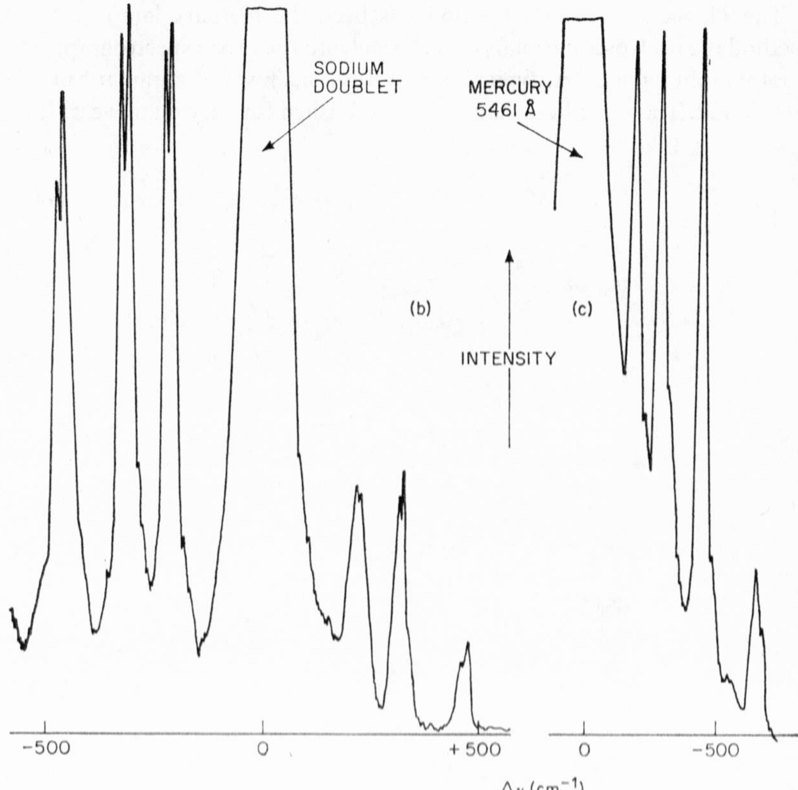

Figure F.1 (contd.) (b) A radiofrequency powered sodium arc and a grating spectrograph were used in this case. Note that each Raman band is a doublet owing to the doublet nature of the source line

(c) A spectrum in the green. Only the Stokes spectrum is shown.

vibrational frequencies, carbon tetrachloride, one sees these properties again demonstrated (source, mercury 4358 Å):

Scattering at:

4514/4509	4449	4419	4405	*4358*	4318	4299	4274	4220/4214 Å

Scattering at:

22150/22178	22481	22626	22722	*22940*	23158	23254	23399	23702/23730 cm^{-1}

Shift, $\Delta\nu(cm^{-1})$, from source:

−762/790	−459	−314	−218	*0*	+218	+314	+459	+762/790

To stress that the important quantity in Raman spectroscopy is the *frequency shift* rather than the absolute frequency of the scattered radiation, we show in Figure F.1 a set of spectra of carbon tetrachloride observed with a number of source lines.

The classic source of radiation has been the mercury lamp, and the method of spectroscopic analysis has been with the prism spectrograph and photographic plate. A typical apparatus is shown diagramatically in Figure F.2. These methods have been established for very many years, and

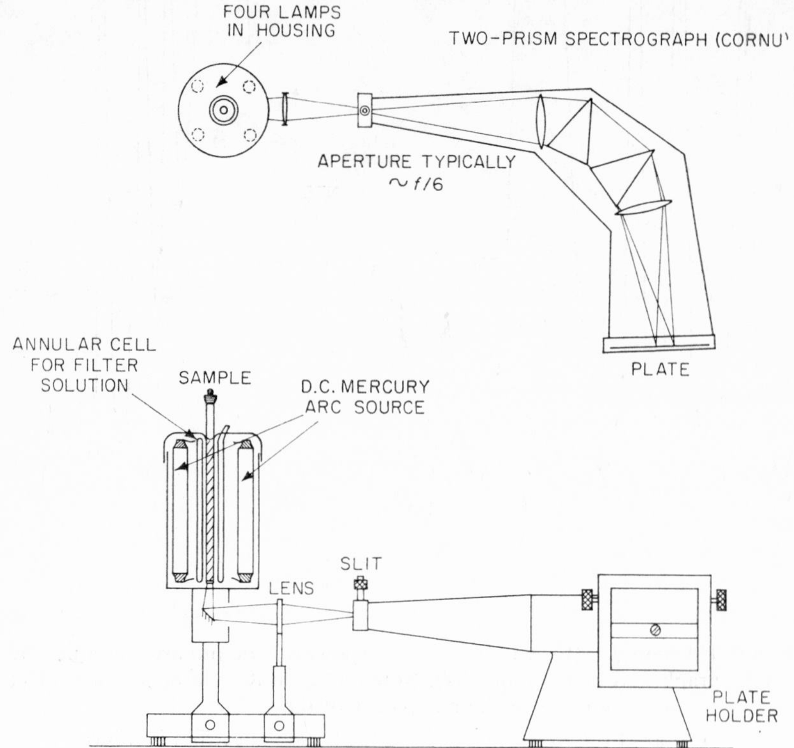

Figure F.2 The classic Raman apparatus, consisting of a powerful source, the sample, and a spectrograph of high dispersion and 'speed'

therefore, although the effect is of incredibly low intensity ($I_{\text{Raman line}} \sim 10^{-8} I_{\text{Source}}$), Raman spectroscopy got off to a very good start immediately after its discovery. By 1939 it was very well established as the most important and practical method for obtaining information on rotational and vibrational transitions. However, it was clear by this time that major practical problems were severely limiting the versatility of Raman spectroscopy. If the sample is coloured, it tends to absorb rather than scatter the radiation. If it is even weakly fluorescent, the emission from this effect will 'swamp' the much weaker Raman bands. Finally, turbidity, dust particles, inhomogeneities, or other optical imperfections in the specimen tend to make the intensity of the source line in the spectrum so high relative to the Raman lines that 'stray light' in the instrument tends to overpower the latter. The development of high-speed detectors for infrared work, during and immediately after the 1939—1945 war, produced conditions ideal for the rapid development of infrared absorption spectroscopy, with the result that the Raman technique dwindled rapidly in importance during 1940—1960. However, the appearance of the 'laser' and its application to Raman spectroscopy has effected a renaissance during the last five years or so, and it is this new 're-vitalized' technique that is now described. Although it is intended that this book should inform (and even convince) committed infrared spectroscopists of the value of the Raman technique, we will conclude this introductory chapter by listing the scope of vibrational and then rotational spectroscopy.

The most familiar use of vibrational spectra is in the 'group frequency' and 'fingerprint' methods of interpretation and qualitative analysis. These hardly require further description but it should be noted that the Raman fingerprint, although *analogous* to its infrared counterpart, is quite different in form for a given compound, because the Raman selection rules are not those for the infrared. In addition, the relative intensities of coincident bands in the two spectra are not the same. This point is nicely demonstrated in Figure 11 of Chapter 2, where infrared and Raman spectra of mesitylene and dicyclohexyl are shown superimposed.

With the advent of relatively complete vibrational spectra (from the addition of Raman data and extended infrared measurements) it has become much more feasible to use vibrational spectra for structural determinations. The vibrational behaviour and the infrared and Raman activity of the modes are very dependent on the geometry of the molecule. Some excellent examples of very simple systems have been collected and listed by H. E. Hallam in a most readable introductory account [*Roy. Inst. Chem. Rev.*, **1**, 39 (1968)]. Unfortunately, there are pitfalls in this method of structural analysis, and the interpretation of the experimental result may degenerate into an art rather than a science. However, the combination of chemical intuition with a sprinkling of good fortune can often lead to 'reliable' structural diagnosis in cases inaccessible by other techniques. A further field of study, for which as much knowledge as possible of the vibrational behaviour of the molecular system is essential, is the analysis of the 'small displacement' forces holding a molecule together. These 'force field calculations' have become far more reliable and useful than they were. Another area where details of all the vibration frequencies of a molecule must be known is in the computation of thermodynamic functions. This stricture particularly applies to the low-frequency modes, which are often difficult to study in the infrared because of the experimental difficulty and cost of operating in the far-infrared. Raman spectrometers *tend* to operate over wider frequency-shift ranges than their infrared counterparts; e.g. a modern Raman spectrometer will cover the range $\Delta\nu = 25-4000\ cm^{-1}$ for most condensed-phase samples, but two infrared spectrometers are required to cover the *equivalent* spectral range (e.g. a grating instrument between 4000 and 200 cm^{-1} and an interferometer from 300 to 25 cm^{-1}).

Turning now to rotational phenomena, these are accessible either by use of microwave absorption (for polar molecules of moderately high mass) or as vibrational–rotational transitions in the infrared. The selection rules for pure rotations to give rise to rotational Raman spectra are much less restrictive than for absorption. All except spherical-top molecules exhibit a spectrum. Unfortunately, it is difficult to record Raman spectra at high resolution. Once they have been recorded, rotational data can be analyzed to yield moments of inertia, and hence interatomic distances of high accuracy, in addition to other physical constants.

1

Introduction

When light passes through any material, some of it is always deflected from the main propagation direction, i.e. it is 'scattered'. The conditions necessary for this to be called Raman scattering will be described in detail later, but briefly, any instantaneous *inelastic* scattering is so called. Most of the scattering is *elastic* (or Rayleigh) scattering. In the Raman effect the original photon energy $h\nu_0$ may be either reduced or augmented, and the (observable) difference is accounted for by an energy transition in the material. When a photon is entirely absorbed or spontaneously emitted, as in normal infrared, ultraviolet, or visible absorption or emission spectroscopy, the radiation is in resonance with the energy level transition of the material during the process. The energy change in Raman scattering is therefore said to be due to 'non-resonant' absorption or emission. The information available about the energy levels of the scattering material is quite different from that obtained from resonance methods. This is partly because the effect is no longer dominated by one transition moment, and partly because the scattered photon brings extra information away with it.

Although vibrational or rotational states are generally involved, electronic states have also been examined by Raman spectroscopy. However, it is important to distinguish Raman scattering from fluorescence. In the latter, resonant absorption occurs but is followed, after a measurable ($\sim 10^{-7}$ sec) lifetime of the excited state, by resonant emission involving either the same lower state (resonance fluorescence) or a different one. Raman scattering and fluorescence are frequently difficult to separate experimentally. In theory also the distinction is not so clear-cut as might be supposed. Conventionally, a Raman process is one in which the lifetime of the excited state is of the order of a vibrational period—about 10^{-12} sec. Fluorescence occurs after a more definite occupancy (perhaps about 10^{-7} sec) of a particular excited state. However, in the resonance

Raman effect (see later) one transition moment dominates the scattering probability, and an arbitrary line has to be drawn between the two time scales, at which resonance Raman passes into short-lived (partially quenched) resonance fluorescence. In practice this region is seldom encountered.

In a Raman spectrometer an intense monochromatic light source is required; experimental and theoretical considerations generally place this somewhere in the visible region. An optical system for collecting scattered light is followed by a highly discriminating monochromator, and a detection system. This system may be modified in the future with the probable advent of a tunable source. At room temperature the Boltzmann distribution factor favours the observation of absorption, in non-resonant as in resonant processes, and therefore the lower-frequency side of the 'exciting line' is generally scanned. This is also the region of fluorescence, called the 'Stokes side' in a phrase often borrowed by Raman spectroscopists. One may be driven to the higher-frequency side ('anti-Stokes') to escape this problem. Figure F.1 includes a Raman spectrum of carbon tetrachloride showing the reduced intensity of the 'anti-Stokes' lines.

From the previous definition of Raman scattering as the concomitant of non-resonant absorption, the 'resonance Raman effect' might seem to be a contradiction in terms. In this case the exciting light is close to an *electronic* absorption band of the sample, i.e. close to resonant absorption in the sample, a situation leading to anomalously high scatter, both elastic and inelastic. Since, in these circumstances, absorption of the scattered radiation is also high, a net gain may not be observed. However, very dilute solutions may exhibit a Raman spectrum under these conditions, when they would not normally do so.

The Raman effect is weak, and the problem is to collect sufficient of the scattered light from a sufficiently large illuminated volume. The necessary volume is minimized as much as possible by devices which focus the exciting light, and sometimes multiply reflect it. Leite *et al.*[1,39] have estimated that the extinction coefficient arising from loss by Rayleigh scattering in pure liquids at 6328 Å is of order 10^{-4} cm^{-1}. The corresponding Raman terms range from approximately 10^{-2} of this, for strong bands, to 10^{-6} (or less) of it for the weakest observable bands. Solutions are correspondingly still less favourable according to dilution. It is generally reckoned that a laser source of at least 10 mW power is necessary for routine Raman spectroscopy.

A Raman spectrum can always be obtained from an ideal sample, but lasers have made it much easier to obtain the spectra of non-ideal samples. For coloured materials the helium–neon and ruby lasers provide an intense stable red source, whereas previous red sources have often been weak, unstable, or 'dirty'. The large majority of coloured samples absorb least at the red end of the spectrum. In the case of powders it is necessary for the excitation to be very intense over a small area; this is easily achieved by focusing a laser beam. The ionized inert-gas lasers and some solid-state lasers can be constructed so as to produce a *useful* power greater than that from any other known continuous monochromatic source. For single-crystal work it is necessary that the electric vector of the exciting light should be in a single defined direction throughout the sample, and this too is easily achieved with a laser (see Appendix 1).

1.1 Electromagnetic Radiation

Before developing the theory of Raman scattering it is necessary to discuss some of the properties of electromagnetic radiation, of which visible light is a small part of the frequency spectrum. Brief mention of lasers is also made, so that their properties as a source of such radiation are apparent.

Electromagnetic radiation is the name given to the process whereby energy is transferred from one oscillating electric dipole to another at a distance, without the intervention of further charge. It is inadvisable to say much more than this without going to the differential (Maxwell) equations of which radiation is the periodic solution, because the concept of cause and effect is inevitably used. Thus, it may be considered that the oscillating electric field produces an oscillating magnetic field at right angles to itself, the oscillations of which produce a new, oscillating, electric field. Magnetic and electric fields are related by their time differentials, and thus a sinusoidal variation in the initial dipole leads to a self-propagating sine wave, all differentials of the sine function having the same form. Any periodic fluctuation may be broken down into the sum of a series of sinusoidal variations, so that any wave behaves as a mixture of sine waves of the appropriate frequencies. However, it must be emphasized that, in the absence of a retarding effect by intervening charges, the electromagnetic wave propagates at the same rate as the fields themselves. This is because no charge accompanies the radiation; i.e. pure fields are involved, capable of separating charges where these exist (in matter), but without effect on a vacuum. The energy per unit area passing a given point, which is proportional to the square of the electric field amplitude, is called the 'intensity' or 'irradiance', although some authors restrict the term 'intensity' to an integrated irradiance over a particular area or solid angle.

In view of its pure field nature, the extraordinary and useful property of electromagnetic radiation is that it can be collimated. A simple oscillating dipole already radiates in preferred directions, energy passing into space according to $(\cos^2\theta)d\omega$, where θ is the angle between the plane perpendicular to the dipole and the given direction, and $d\omega$ is a solid angle increment. Passage of part of this radiation through a suitable slit and lens (or mirror) system produces a 'beam' with well defined direction and position. Alternatively, in the laser a resonant cavity is used to cause selective enhancement of the required part of the radiation from a series of oscillating dipoles (emitting atoms). This process depends on the quantum mechanical phenomenon of stimulated emission, in which part of the energy of an emitter is proportional to the radiation density surrounding it, and appears with the same directional properties as the stimulation.

The directional and spatial properties of light are often summarized in the phrase 'light travels in straight lines'. It would be more accurate to state that light may be confined to a single direction in space. This distinguishes it from static field phenomena, which normally cannot be prevented from 'radiating' in all directions from a point 'source'.

These properties of electromagnetic radiation are easily acceptable in the quantum formulation. The quantum of electromagnetic radiation is called the 'photon', with familiar energy

$$E = h\nu \tag{1}$$

where h is a universal constant (Planck's constant) and ν is the associated frequency. It seems reasonable that particles should have the properties described. These properties also follow directly from the wave formulation. There is no need to show the mathematical details but, in the wave analysis of a beam of radiation, the spatial confinement results from the destructive interference of all but the 'secondary wavelets' travelling in the direction of the 'primary wavelets' from which they originated. This topic will be taken up again when the interaction with matter is considered.

1.1.1 *Coherence and Polarization*

The electromagnetic wave discussed above 'consists' of an oscillating electric field and an oscillating magnetic field instantaneously at right angles to one another, both being at right angles to the direction of propagation. At a particular point in space all the attributes of the wave at that point (except the 'sense' of the propagation) may be characterized by reference to the 'electric vector' or the magnitude, direction, and time-variance of the electric field at the point. In the most general case, the tip of an arrow representing the electric vector traces out an irregular shape at an irregular rate; the shape may approximate to an ellipse. Clearly, unless this ellipse remains approximately constant in shape, size, and orientation, and is traced out with an approximately constant frequency, it is not very helpful to know its short-term details. Similarly, unless the ellipse characterizing the field at a neighbouring point bears some relationship to the first, it does not form a very adequate description of a real beam of radiation.

If all such ellipses characterizing a beam of radiation have the same eccentricity, orientation, and handedness, and, more important, if these properties are time-invariant, the radiation is said to be 'polarized'. If, in addition, all the ellipses on a surface perpendicular to the propagation direction of the beam are in phase, it is said to be 'spatially coherent'. The distance apart over which the ellipses on two such surfaces bear a definite relationship to one another is the longitudinal or time-coherence. Two or more beams (or sources) are said to be coherent if the ellipses of one bear a definite relationship to those of the other over a period of time. The coherence of two beams is an important factor in considering their combination. If they are incoherent, it is sufficient to sum the two intensities (energies) to obtain a correct result. If they are coherent, the electric vectors must be summed by vector addition with due regard to spatial and time variance, before taking the square (to which intensity is proportional).

Coherence is not important in the normal Raman effect, but it is mentioned here because of its relationship to monochromaticity and polarization. An ideally monochromatic beam of radiation is one in which the frequency is identical at all points in space and time. If, in addition, the intensity is to have only one maximum through the region of the beam, it is also necessarily polarized and coherent. Briefly, one states that a perfectly monochromatic wave is polarized and coherent. A real beam of radiation departs from this ideal in various ways depending on its method of production. The output of a laser is much closer to the ideal than any other yet obtained in the visible region. Unfortunately a normal laser cavity resonates at several frequencies falling within the width of the fluorescence band concerned; it is possible to suppress the side-bands for special purposes.

When the ellipse degenerates into a straight line, the radiation is said to be 'plane-polarized', and the polarization no longer has a handedness. It is this type of polarization which is most useful in the Raman effect, and our subsequent discussions will be confined to it. The electric vector is confined to a single direction in space, and the Raman effect, along with most effects of electromagnetic radiation, arises from interaction of the electric vector with matter. Unfortunately, before the significance of polarization was understood, a convention was made whereby certain phenomena were linked with a 'plane of polarization' of polarized light, afterwards discovered to be the plane of the magnetic vector. In this book the term is therefore avoided, and reference will be made only to the direction of the electric vector.

Experimentally it is easy to produce plane-polarized light, e.g. with the familiar Polaroid sheet or Nicol prism polarizers. As described in Appendix 1, continuous gas lasers are generally built in a way which produces a plane-polarized output. This is by use of the 'Brewster angle' phenomenon whereby light incident on the surface of a transparent medium at a certain angle is reflected only in one polarization and perfectly refracted in the other. The windows on the ends of the 'plasma tube' of the gas laser are set at this angle, so that laser action can only occur in this one polarization. The totally refracting (and therefore transmitting) polarization is that in which the electric vector is in the plane formed by the incident and refracted rays. It is important to remember this when refracting polarized light in prisms, since the effect falls off only slowly for angles away from the Brewster angle. Gratings also show a dependence of transmission on angle and hence on wavelength.

1.1.1.1 *Retardation plates.* 'Half-wave' plates are commonly used to turn the direction of polarization of a laser beam. They are a special case of the retardation plate, which we discuss now.

When light strikes the surface of an anisotropic material, it is generally split into two plane-polarized components, with a 90° angle between the two electric vectors. This is the phenomenon of birefringence, to which we shall return later in connection with single-crystal studies. Here we need note only that the two beams are subject to a different refractive index, i.e. they propagate at different velocities. For this reason they do not normally interfere inside the material but, if a very thin specimen is used so that the phase difference on emerging is only a fraction of a wavelength, the two components then recombine coherently with interesting effects on the polarization of the radiation. This is the principle of the retardation plate, and we will examine the mathematics of the effect here as a useful introduction to this type of calculation in optics.

We will consider only the effect on initially plane-polarized light. Immediately before entry, the electric vector behaves according to

$$|e| = a\cos(\omega t) \qquad (2)$$

where a is the amplitude and ω the angular frequency. Immediately after entry into the plate, the vector is resolved into a 'fast' component

$$|e_f| = a\cos(\omega t)\cos\theta \tag{3}$$

along the 'fast' axis, and a 'slow' component

$$|e_s| = a\cos(\omega t)\sin\theta \tag{4}$$

along the 'slow' axis, where θ is the angle between the initial direction of the electric vector and the 'fast' axis. Let us re-define the phase angle to be zero for the 'fast' beam on emergence. Then, at this point,

$$|e_f| = a\cos(\omega t)\cos\theta \tag{5}$$

$$|e_s| = a\cos(\omega t - 2\pi r)\sin\theta \tag{6}$$

for an 'r wave' retardation plate.

The vectors e_f and e_s now recombine according to the principle of vector addition, and we define ϕ as the instantaneous angle between the initial direction of the electric vector, and that of the recombined vector, e_r. Then

$$|e_r| = a\cos(\omega t)\cos\theta\cos(\theta+\phi) + a\cos(\omega t - 2\pi r)\sin\theta\sin(\theta+\phi) \tag{7}$$

whilst ϕ itself is found from the cancellation of the opposing components:

$$a\cos(\omega t)\cos\theta\sin(\theta+\phi) = a\cos(\omega t - 2\pi r)\sin\theta\cos(\theta+\phi) \tag{8}$$

If we expand the term containing the retarded phase:

$$\cos(\omega t - 2\pi r) = \cos(\omega t)\cos(2\pi r) + \sin(\omega t)\sin(2\pi r) \tag{9}$$

it is immediately obvious that for

$$r = \tfrac{1}{2}n$$

with n any integer, the time-dependent terms cancel from the equation for ϕ, so that ϕ is a function only of θ, the orientation of the retardation plate. This implies that the light emerging is still plane-polarized, and we further find

$$\phi = 0,\ n \text{ even} \tag{10}$$

$$\phi = \pm 2\theta,\ n \text{ odd (sign depends on quadrant)} \tag{11}$$

Making n odd, but as small as possible to retain coherence, we see that a 'half-wave plate' may be used to turn the plane of polarization of plane-polarized radiation. This device is important in the use of the laser source since it provides a means of setting the electric vector of the exciting radiation in any desired direction. Another device used in optics is the quarter-wave plate, which has a more complex action. For example, with $\theta = \frac{1}{4}\pi$ (only):

$$\phi = \omega t - \tfrac{1}{4}\pi \tag{12}$$

which corresponds to circularly polarized light. If detected 'incoherently' this is effectively depolarized.

It should be noted that more powerful methods exist for analyzing these problems using more advanced mathematical techniques, and these are to be found in most good optics textbooks.

1.2 Interaction with Matter: Rayleigh Scatter

The interaction of radiation with matter is a vast subject, and one in which the approach changes from one theoretical physicist to another. One may find purely quantum formulations, purely classical formulations, and desperate attempts to extract something useful from judicious combinations. All we shall attempt is to give a feeling for the ideas involved, since even a limited insight is usually more help than a bare statement of observed facts.

It is usual to consider two broad categories of interaction—scattering and resonance. A single theory of scattering suffices at least to prepare the ground for such diverse phenomena as optical and X-ray diffraction, refraction, and Rayleigh, Tyndall, and Raman scattering. The formalism likewise extends to fields other than the electromagnetic, covering, for example, electrons and thermal neutrons. Resonance has already been mentioned.

In the classical model, it is considered that matter contains separable charges held together by forces additional to purely coulomb attraction but not further specified. The arrival of a wave at the boundary with matter sets up an oscillating separation of these charges, i.e. an oscillating electric dipole. This dipole radiates at the frequency of oscillation, and this radiation is, in the general sense, scatter. Scattering by widely separated particles of much smaller dimensions than the wavelength of the radiation (as with visible light in a rarefied gas) is known as 'single scattering', the criterion being that the scatter from one particle is insufficient to affect the scatter from the next. In this case the greater part of a beam of radiation passes straight through the sample, and a small proportion is scattered in other directions. We consider that the particle has a 'polarizability' α, connecting the applied field with the resultant dipole μ according to

$$\mu = \alpha e = \alpha a \cos(\omega t) \tag{13}$$

where $e = a \cos(\omega t)$ is the applied field.

In general the polarizability varies with the frequency of the field, so it must not be expected to be the same for light as for a static field. The dipole radiates according to the classical laws, and if the initial radiation is polarized the scatter may also be polarized, so long as the particles are isotropic with μ and e coincident. We shall return to this point later. This is Rayleigh scattering, and its intensity is proportional to the mean square value of μ.

1.2.1 *Raman Scatter*

It may be that the irradiated particle is itself undergoing some sort of periodic fluctuation (perhaps thermally excited) quite distinct from the forced oscillatory charge separation already discussed. Consider, for example, the vibration of a diatomic molecule. Clearly this will affect the polarizability, which becomes

$$\alpha[1+b\cos(\omega_v t)] \tag{14}$$

in the harmonic oscillator approximation, where ω_v is the angular vibrational frequency. Thus

$$\mu = \alpha a \cos(\omega t) + \alpha a b \cos(\omega t)\cos(\omega_v t) \tag{15}$$

The second term may be broken down to give

$$\tfrac{1}{2}\alpha a b\{\cos[t(\omega+\omega_v)]+\cos[t(\omega-\omega_v)]\} \tag{16}$$

and so it is that these two ('beat') frequencies appear weakly in the scattered radiation. This is the phenomenon of Raman scattering. The vibrations of polyatomic molecules in general give rise to Raman scattering, although selection rules apply where the molecule has any symmetry. Other Raman scattering phenomena, particularly the rotational and electronic Raman effects, are more easily understood as quantum mechanical phenomena.

Quantization affects the picture in various ways. In single Rayleigh scattering we consider occasional collisions between photons and particles, but the likelihood of a collision is governed by a term closely akin to the classical polarizability. Collision is followed by scattering of the whole photon, in a direction and with a polarization giving on average the classical distribution. Quantum mechanics also gives us an insight into the particle 'polarizability' (as we shall continue to call it). Since the electrons are much lighter than the nuclei, in the atoms of which we now consider the particles to be composed, we assume that the polarizability arises from distortions of the electron charge cloud. Quantum mechanically we consider such distortions or perturbations to arise as perturbations of the electronic wave function by the addition of small amounts of all the other possible electronic wave functions of the system. Each such perturbing wave function contributes to the overall probability of photon scatter according to the product

$$M_{nm} M_{mn} \tag{17}$$

where M_{nm} is the transition moment for the transition n→m, n being the existing state and ψ_m the perturbing wave function under consideration. The state 'm' is called a 'virtual state' or a 'virtual level', and it must not be supposed that the particle in any sense attains it; its role is simply to perturb the existing state.

It is clear that, if we consider quantization in relation to Raman scattering, incident and scattered light have a different photon energy. This energy difference is just the quantum of vibrational energy concerned. Herein lies the important difference from the classical analysis. Classically it is considered that the molecule is already vibrating, and continues with the same amplitude after Raman scattering occurs. Quantum mechanically we see that the incident radiation actually promotes or demotes the vibrational energy level. This is in accord with the observed fact that the 'anti-Stokes' Raman lines (appearing by the enhancement of photon energy by vibrational quanta) fall off in intensity with increasing shift, according to the predicted thermal occupation of the levels (see Figure F.1). There is, of course, no contradiction here of the classical theory; one may add to the classical arguments the quantum hypothesis of moment of scattering energy exchange between the radiation and the matter, without otherwise affecting them. There is, however, no *classical* argument to justify doing so.

The products (17) for Rayleigh scattering are replaced by

$$M_{nm} M_{mi} \tag{18}$$

for Raman scattering, where 'i' is the final state of the particle after scattering. Classically, the intensity of Raman scattering is related to the constant b used in equation (14). Clearly,

$$b = c(\partial\alpha/\partial q_i) \tag{19}$$

where c is the amplitude of the vibrational disturbance, and q_i is the ith normal coordinate; this relation holds for an infinitesimal amplitude of vibration but is a reasonable approximation for real amplitudes. As we

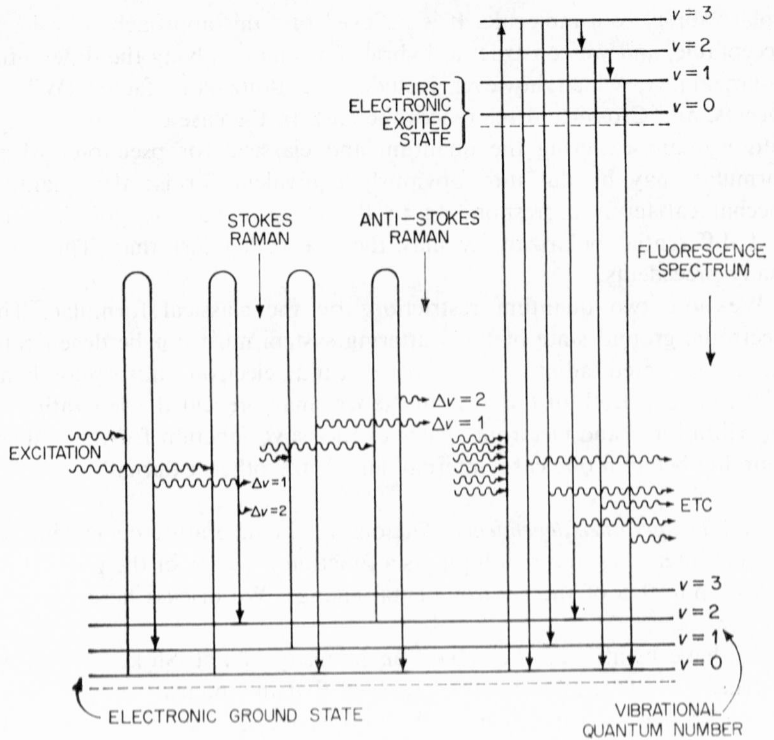

Figure 1.1 Stokes and anti-Stokes Raman and fluorescence processes.

have seen, there is doubt as to the classical value of c, but its average may be taken as a bulk constant at a particular temperature, so that the differential polarizability is the important quantity controlling Raman intensities.

Although the relationship of the quantum formulation to the classical formulation is not immediately obvious, the important result is that the full expression may still be written in terms of a 'Raman tensor' controlling the probability of Raman scatter. The fact that the differential polarizability is also a tensor quantity is a point which we shall discuss more fully later, and the same selection rules may be shown to apply to it. Further, the quantum formulation is of little value in predicting intensities, since the transition moments are not generally known. However, we find that the observed intensities are not inconsistent with some rather naive quasi-classical ideas which we apply to construct a model of the differential polarizability of a molecule. It is believed that this approach is basically acceptable, and we construct a 'hybrid' formula involving the differential polarizability, which however includes the Boltzmann factor (Wilson, Decius, and Cross[2] and references therein). In the case of both Rayleigh and Raman scattering the quantum and classical (or pseudo-classical) formulae may be put into obviously equivalent forms; the quantum mechanical terms corresponding to the classical particle polarizability and differential polarizability may then be called just that. There are many precedents.

We note two quantum restrictions on the classical formulae. The electronic ground state of the scattering system must not be degenerate, and the scattered radiation must not be near an electronic absorption band of the system. Both of these circumstances may prevent the separation of the vibrational and electronic parts of the wave function for the system, coupling being in one case 'internal' and in the other 'external'.

1.2.1.1 *Temperature dependence*. The observed temperature-dependence of Raman intensities in a real liquid is a function not only of the population factors but also of the environmental change. We discuss here only the former.

We have mentioned the fall-off in intensity of anti-Stokes lines with increasing shift. It must not be imagined that the intensity appearing on the anti-Stokes side is lost from the Stokes side. The transition from the first excited vibrational level to the second is, in fact, more favourable than the fundamental 0→1 transition in the Raman as it is in the infrared, so a temperature enhancement of Stokes lines also occurs. The anharmonicity of the vibration means, however, that the transition 1—2 is at a slightly different frequency from 0—1. These 'hot' bands may even be resolvable in highly anharmonic vibrations.

Neglecting environmental changes and the possible resolution of 'hot' bands, the intensity ratio for two absolute temperatures T_1 and T_2 on the Stokes side is given[3] by

$$I_{T_1}/I_{T_2} = (1-e^{-h\nu_i/kT_2})/(1-e^{-h\nu_i/kT_1}) \tag{20}$$

or on the anti-Stokes side by

$$I_{T_1}/I_{T_2} = e^{-h\nu_i/kT_1}/e^{-h\nu_i/kT_2} \tag{21}$$

The ratio of anti-Stokes to Stokes intensity is, however:[3]

$$I_{as}/I_s = (1-e^{-h\nu_i/kT})e^{-h\nu_i/kT}\left(\frac{\nu+\nu_i}{\nu-\nu_i}\right)^4 \tag{22}$$

expressing the fact that scattered intensity is proportional to the fourth power of the absolute frequency.

1.2.2 *The Resonance Denominator*

At the outset it was stated that we were treating scattering and resonance as separate problems. Of course they are not really separate, but all we have said so far does depend on the *absence* of resonance. That is, the radiation which we consider as interacting with the matter must be far removed in frequency from any dipole-active resonant frequency associated with the matter. If such a resonant frequency is approached, the induced dipole becomes abnormally large. This is accompanied by classically continuous but actually discrete absorption; however, this does not prevent the induced dipole from producing anomalously large scatter. The scattered radiation suffers absorption in its turn, so the characteristic of *resonance* scattering is that it can be observed at very low concentrations of the scattering species, rather than its resultant magnitude at higher concentrations.

In the quantum formulation, the Rayleigh scattering term associated with the virtual level m is more correctly written

$$M_{nm} M_{mn} \left(\frac{1}{\nu_{mn} - \nu} + \frac{1}{\nu_{mn} + \nu} \right) \tag{23}$$

where ν_{mn} is the frequency associated with the transition m→n, and ν is the frequency of the incident radiation. This expression ignores the 'damping' term which prevents it from increasing to infinity as ν approaches ν_{mn}. However, it does become very large in these circumstances, causing one virtual level to dominate the perturbation, so that it alone need be considered. The probability of a photon collision becomes very high indeed, and although many such collisions result simply in the transition n→m, according to a second (absorption probability) term, the scattering is also drastically increased according to the expression given. It will be clear that if the incident radiation actually contains ν_{nm}, the distinction between resonance scattering and resonance fluorescence is not altogether clear-cut.

The corresponding Raman scattering terms are

$$M_{nm} M_{mi} \left(\frac{1}{\nu_{mn} - \nu} + \frac{1}{\nu_{mi} + \nu} \right) \tag{24}$$

similar enhancement being expected. For further details of the resonance Raman effect, see refs. 4 and 5. There is little doubt, however, that much of the phenomenal success in observing the Raman spectra of yellow and red samples, using 6328 Å He–Ne laser excitation, is due to a fortunate balance in which resonance enhancement far exceeds absorption. This is often referred to as the region of 'preresonance'.

1.2.3 *Optically Dense Materials*

All the preceding discussion has been confined to single scattering. It is clear that this condition cannot hold for *optically dense* materials, in which there are a large number of scattering particles within a cube of dimensions equal to the irradiating wavelength. Even a gas at s.t.p. is optically dense for visible light, since it contains some 10^6 molecules in such a cube. Raman scattering is so weak that the results are not affected for normal exciting fields, but in Rayleigh scattering the scatter soon reaches a level at which secondary scatter is important, and the whole scattering system must be treated together. Although this has little relevance to ordinary Raman scattering, it would not be satisfactory to leave the reader with this incomplete picture of interaction.

The results for some simple systems are not too different for the classical and quantum approaches. This is essentially a wave problem, and either classical or quantum mechanical scattering terms may be incorporated into the same wave solution. Rayleigh scattering is coherent (Raman scattering normally is not coherent), and it is also coherent with the exciting radiation having a fixed phase difference with it. The problem of arriving at a solution simultaneously for all the particles present, taking into account the 'coupling' or secondary scattering, is quite formidable. A solution, the Ewald–Oseen theory, has been given, and its results in fact simply reproduce the observations. Inside the medium the incident wave is completely extinguished (the Oseen extinction theorem), and the coherent superposition of incident and scattered radiation gives a refracted wave of wavelength λ/n and propagation velocity c/n, with direction according to Snell's law, n being the refractive index. Outside the medium a reflected wave appears.

This theory assumes that the scattering particles are equally spaced. If the statistical fluctuations of, say, the kinetic theory of gases are introduced, a weak general scattering also occurs. Further, it appears that, so long as the refractive index is close to unity in an ideal gas, the same weak general scatter is predicted either as a result of single scattering or by the Ewald–Oseen treatment. Thus, what we termed Rayleigh scatter in rarefied media is continuous with the scatter due to density fluctuations in optically dense media. Accordingly, the latter may be called Rayleigh scattering (historically it received the name first, being more obvious than the former experimentally), and we are invited to ignore coupling and refraction, as a first approximation, when dealing with it.

These results also suggest a more amenable treatment of optically dense media. The original differential equations (the Maxwell equations), which have electromagnetic radiation as one of their solutions, may be written in a 'macroscopic' way which takes account of the presence of a uniform or continuous density of electric charge. The new solutions again match with experience, and again Rayleigh scattering appears as a result of fluctuations in density.

1.2.4 *Stimulated Raman Scattering*

It has already been remarked that Raman scattering is weak enough to be treated as 'single' in most cases. The scattering is also normally incoherent, so the special effects comparable to refraction could not appear. However, if the excitation used is sufficiently intense, e.g. from giant pulsed solid-state lasers, a quantum phenomenon related to normal stimulated emission is encountered; gain occurs, and the scattering becomes coherent and is emitted in fixed directions. A significant proportion of the exciting radiation may be converted into Raman radiation, which usually appears at the most intense normal Raman frequency. Naturally this stimulated emission acts in turn as exciting radiation, so that in general frequencies ($\nu \pm n\nu_i$) appear, the intensity falling off with n (although not so quickly as would appear from this simplified explanation). It may be surmised that stimulated Raman emission is accompanied by ordinary Raman emission arising from density fluctuations. No chemical applications of this effect have yet appeared, since only one or two Raman shifts are normally observed, but it is a useful means of producing coherent light, slightly shifted in frequency from a particular laser output. It is in some ways unfortunate that this effect dominates other interesting phenomena, which may be observed with very intense excitation. Such experiments must always be performed with illumination below the threshold for the stimulated effect.

1.2.5 *Hyperpolarizability*

Very intense irradiation has no effect on the normal Raman spectrum below the threshold for the stimulated Raman effect. However, interesting phenomena appear in the region of twice the incident frequency. This is because account must be taken, in classical terms, of the 'hyperpolarizability' of the scatterer. If the excitation is sufficiently intense, the polarization of the scatterer is no longer adequately represented by equation (13):

$$\mu = \alpha e = \alpha a \cos(\omega t)$$

but by

$$\mu = \alpha e + \tfrac{1}{2}\beta e^2 + (1/6)\gamma e^3 + \ldots \qquad (25)$$

where β and γ are the first and second hyperpolarizabilities. This deviation becomes noticeable when μ is of the same order of magnitude as permanent molecular dipoles. It may be seen that this leads to generation of harmonics in Rayleigh scattering (light of frequency 2ν, 3ν, . . .). The derivatives $\partial\beta/\partial q_i$ of β with respect to vibrations and rotations may also be non-zero, leading to the hyper-Raman effect, or scattering at $(2\nu \pm \nu_i)$. The symmetry properties of $\partial\beta/\partial q_i$ are quite different from those of $\partial\alpha/\partial q_i$, so quite different modes may be active in the hyper-Raman, as compared with the normal Raman effect. It often happens that infrared and Raman 'inactive' modes are active in the hyper-Raman effect, so this should in principle be a way of obtaining the frequencies of these hitherto 'silent' modes, and completing the vibrational information for highly symmetric molecules.

1.2.6 *Multiphoton Effects*

In the quantum formulation of scattering phenomena, there are distinct coefficients describing the probability of multiparticle processes, in which two or more particles undergo simultaneous scattering at the same site. Such processes are of course rare, and in some cases they merely augment the single-particle phenomena by an insignificant amount. The reader will be familiar with the related molecular phenomena of combination, overtone, difference, and 'hot' bands in infrared spectroscopy. In Raman spectroscopy the latter phenomena appear (see Section 1.2.7), but there is also the possibility of 'multiphoton' processes. The hyper-Raman effect appears as such a process in the quantum formulation

$$h\nu + h\nu \rightarrow h(2\nu \mp \nu_i) \pm h\nu_i$$

where $(2\nu \pm \nu_i)$ are the hyper-Raman frequencies, and $h\nu_i$ appears in or disappears from a molecular energy level.

Another possible two-photon process is

$$h\nu_1 + h\nu_2 \rightarrow h(\nu_2 - \nu_1) + 2h\nu_1 \qquad (26)$$

where the energy $h(\nu_2 - \nu_1) = h\nu_i$ appears in a molecular energy level. This is the inverse Raman effect, so-called because it implies absorption of the photon $h\nu_2$ in the presence of excitation $h\nu_1$, with the same selection rules for the molecular energy levels as in the normal Raman effect. The term 'Stoicheff absorption' has also been used. In principle a sample irradiated with light of frequency ν_1 (at an intensity below the threshold for the stimulated Raman effect) is able to absorb, from an intense continuum irradiating it at the same time, all the frequencies ν_2, ν_3, . . . corresponding to Raman shifts from ν_1. In practice such a continuum is difficult to achieve, and the effect has been observed only by a few specialist investigators.

Note that *excitation* of molecular levels occurs with $\nu_2 > \nu_1$, so that the favourable Boltzmann factor on the Stokes side in the normal Raman effect appears on the anti-Stokes side in the inverse effect.

The stimulated, hyper- and inverse-Raman effects are all examined in more detail in Chapter 8 where experimental details are also given. We note here that specialist equipment is necessary for their observation, since no commercially available spectrometer is directly suitable.

1.2.7 *Overtone and Combination Bands*

Classically these have a slight activity in the Raman effect, as in the infrared, arising from the anharmonicity of the vibrations concerned and non-zero values for

$$\partial^2\alpha/\partial q_i^2 \text{ and } (\partial/\partial q_j)(\partial\alpha/\partial q_i)$$

The reader may verify that these result in the presence of dipoles of frequency $(\nu \pm 2\nu_i)$ and $(\nu \pm \nu_i \pm \nu_j)$, which radiate in the normal way. It is also sometimes possible to detect weak Raman lines, excited by light which itself originates as Raman scattering.

1.2.8 *Selection Rules*

From consideration of the classical molecular polarizability, we have previously written a simple equation (14):

$$\mu = \alpha E$$

where we temporarily use a static field E for purposes of illustration. μ and E are vector quantities, and since we did not specify their directions the above equation is strictly written

$$|\boldsymbol{\mu}| = \bar{\alpha}|\boldsymbol{E}| \tag{27}$$

The directions of $\boldsymbol{\mu}$ and $\boldsymbol{E}$ do not necessarily coincide, and for a full specification of the system we must consider the three components of each of them on the axes of a fixed cartesian coordinate system. Then we write

$$\boldsymbol{\mu}_x = \alpha_{xx}\boldsymbol{E}_x + \alpha_{xy}\boldsymbol{E}_y + \alpha_{xz}\boldsymbol{E}_z \tag{28}$$

$$\boldsymbol{\mu}_y = \alpha_{yx}\boldsymbol{E}_x + \alpha_{yy}\boldsymbol{E}_y + \alpha_{yz}\boldsymbol{E}_z \tag{29}$$

$$\boldsymbol{\mu}_z = \alpha_{zx}\boldsymbol{E}_x + \alpha_{zy}\boldsymbol{E}_y + \alpha_{zz}\boldsymbol{E}_z \tag{30}$$

The various α_{ij} are the nine 'components' of the 'polarizability tensor'. Normally $\alpha_{ij} = \alpha_{ji}$, so there are only six independent components. A more succinct notation has been developed to express this type of relationship. We denote the array or 'matrix'

$$\begin{pmatrix} |\mu_x| \\ |\mu_y| \\ |\mu_z| \end{pmatrix}$$

by the symbol $\boldsymbol{\mu}$. $\boldsymbol{E}$ is similarly defined, whilst

$$\boldsymbol{\alpha} = \begin{pmatrix} \alpha_{xx} & \alpha_{xy} & \alpha_{xz} \\ \alpha_{yx} & \alpha_{yy} & \alpha_{yz} \\ \alpha_{zx} & \alpha_{zy} & \alpha_{zz} \end{pmatrix} \tag{31}$$

The special brackets enclosing the arrays or matrices are used to indicate that this is what they are. We may now write a matrix equation

$$\boldsymbol{\mu} = \boldsymbol{\alpha}\boldsymbol{E} \tag{32}$$

which replaces the three previous linear equations, matrix multiplication being defined in such a way that the same result is implied by either notation. A whole self-consistent matrix algebra exists, and numerous standard texts are available for the reader who is unfamiliar with these ideas.

The linear polarizability $\boldsymbol{\alpha}$ is called a two-dimensional or second-rank 'tensor'. The name arises from the related mechanical problem of stress and strain. The first and second hyperpolarizabilities [β and γ of eqn. (25)] are, respectively, third- and fourth-rank tensors.

Any or all of the nine components of the linear polarizability tensor may be modulated by molecular vibrations or rotations. Each of the components may be assigned to one of the symmetry types of the molecular point-group, and the selection rule is usually that a vibration or rotation modulates only those components having the same symmetry as itself. The modulated components are of course those which make the vibration Raman-active. This rule extends to overtone and combination bands, and to the modulation of the components of the hyperpolarizability in the hyper-Raman affect. The symmetry classifications for a symmetrical second-rank (Raman) and third-rank (hyper-Raman) tensor are given in an Appendix. For difference bands and ‘hot’ bands, i.e. vibrational transitions in systems not in the vibrational ground state, the symmetry of the transition is not that of the final state, but rather that of the product of upper and lower states. However, if such transitions are formulated[6] as $(\nu_k - \nu_i)$ and $(\nu_k + \nu_i - \nu_i)$ respectively, the selection rules are the same as for $(\nu_k + \nu_i)$ and $(\nu_k + 2\nu_i)$.

The description of the polarizability varies with the coordinate system chosen to describe it. It will be appreciated that the selection rules take the simplest form when the three axes are chosen as important molecular directions. Polarizability components are often assigned to symmetry types (irreducible representations) in character tables; here the axes are chosen to coincide with important symmetry elements. It is most important for single-crystal work to know which axis has been chosen to coincide with each symmetry element, and much confusion can arise by assuming that a different choice has been made. For safety one should always use character tables in which the choice is explicitly stated, and always check that such a choice is being used. Alternatively, the reference frame may be *rotated* to any desired position. In matrix notation, if the relationship between the (orthogonal) systems is

$$\mathbf{x}' = \mathbf{T}\mathbf{x} \tag{33}$$

then

$$\boldsymbol{\alpha}' = \mathbf{T}\boldsymbol{\alpha}\mathbf{T}^{t} \tag{34}$$

where $\mathbf{T}^{t}$ is the ‘transpose’ of $\mathbf{T}$. Readers desiring to perform such rotations are strongly advised to become familiar with matrix methods. However, short-cuts of the following sort may be useful. Consider a rotation of 45° about z:

$$x' = x + y \tag{35}$$

$$y' = x - y \tag{36}$$

Then

$$x^2 - y^2 = (x+y)(x-y) = x'y'$$

so

$$\alpha_{xx} - \alpha_{yy} = \alpha_{x'y'} \tag{37}$$

However, attempts should not be made to derive normalization factors in this way.

It is important to realize the physical significance of Raman scattering by virtue of modulation of a particular component, say α_{ij}. Remember that this component is concerned with an exciting electric field along j, and a resultant dipole along i; the dipole is modulated in the same way as the component, so that light at the Raman shifted frequency is emitted *as from a dipole of direction i.* The Raman intensity is therefore at a maximum in the plane perpendicular to i, falls off according to the square of the direction cosine with that plane, and is polarized in the plane containing i and the propagation direction. This result is of fundamental importance.

The quantum mechanical formulation of selection rules and light polarization is in every way analogous to the classical. Scattering is controlled by the 'Raman tensor', and as a tensor quantity this is subject to the same symmetry considerations as $\partial\alpha/\partial q_i$. The light distribution (after averaging over many photons) and polarization are the same for a non-zero value of R_{ij}, the ij component of the Raman tensor, as for a modulated polarizability component α_{ij}. Under normal circumstances, also

$$\alpha_{ij} = \alpha_{ji} \tag{38}$$

$$\partial\alpha_{ij}/\partial q_i = \partial\alpha_{ji}/\partial q_i \tag{39}$$

$$R_{ij} = R_{ji} \tag{40}$$

so that there are only six independent components of the Raman tensor to consider.

This is as far as we wish to take the theory of Raman scattering before considering some of the available experimental geometries; these are discussed in Chapter 2. In Chapter 3 we develop the application of the theory to fluids, with regard to both intensities and depolarization ratios. In Chapter 4 we consider single-crystal spectroscopy. It will be apparent already that in the case of fluids much information is lost because the molecules are randomly oriented. As we shall see, much may also be salvaged. In crystals orientation is normally fixed, but the molecules (insofar as they exist as independent entities in a crystal) are subject to crystal-field distortion. Useful information is also obtainable from non-molecular crystals.

2

Experimental

In this chapter the basic requirements of a modern Raman spectrometer and the value of the laser source will be discussed. A comprehensive survey of the commercially available spectrometers is also included, with particular reference to sampling arrangements, versatility, and value to chemists. Some of the Raman data in the literature have been recorded using special sampling techniques not really applicable to more general problems (spectra of melts, samples under very high pressures, etc.), and these experiments are described in the chapters relevant to the results and not in this chapter. The experimental arrangements applicable to gases are also discussed in Chapter 5.

2.1. Introduction

In the basic Raman spectrometer, the sample under examination is subjected to irradiation from a suitable monochromatic source, and the Raman spectrum is observed by use of a system comprising a monochromator, detector, and recorder.

Before the advent of laser sources, the most popular device for producing intense monochromatic radiation was the low-pressure mercury discharge lamp. This device produces a number of intense lines of which those at 4358 and 5461 Å are particularly useful. To prevent interference from fluorescence, spectra from shorter-wavelength source-lines, and (in the case of grating instruments) high-order spectra, filters were always used; e.g. concentrated aqueous sodium nitrite for 4358 Å, and concentrated aqueous potassium chromate for 5461 Å. The major problems with the early discharge sources were linewidth and continuum between the intense discharge lines, but these were largely eliminated by keeping the pressure of mercury vapour in the lamp to a minimum. The Toronto Arc had water-cooled mercury electrodes separated by 1—2 metres of glass tubing coiled around the sample, but it absorbs enormous quantities of power (typically ~2—5 kW) and therefore dissipates a tremendous amount of heat; cooling of the sample is therefore essential. To examine deeply coloured specimens, numerous sources have been devised, of which the helium discharge and rubidium radiofrequency powered electrodeless arc are probably the most successful. Both of these sources have lines of high intensity in the red or near-infrared. Unfortunately it is difficult to detect radiation at long wavelengths, and for these experiments, since photoelectric detectors are relatively insensitive, photographic methods and long exposures were normal.

Numerous sampling arrangements were devised to minimize the intensity of the source line entering the spectograph, and to enable as much Raman light as possible to be collected. In Figure 2.1 are illustrated some successful Raman sources and sample arrangements which have been widely used.

The detection of the very weak Raman spectrum has always been a problem. Most methods make use of a specially designed spectrograph such as the Hilger double glass prism instrument in which an aperture of ~*f*/5 is available and photographic quarter-plates are used for recording. Although successful for work with the blue sources, the arrangement presents difficulties when working at longer wavelengths because the photographic plates are insensitive and the instrument dispersion is low. When using red sources a wide-aperture grating spectrograph is normally used to overcome this limitation. The insensitivity of the plates necessitates long exposures (sometimes up to 50 hours), and this fact has tarnished the reputation of Raman spectroscopy as a routine technique.

Photoelectric recording spectrometers have been available (e.g. from Cary, and Hilger and Watts) but the importance of the technique in chemistry has just not compared with its "competitor" infrared spectroscopy. There has been no Raman instrument equivalent to the small low-priced versatile infrared spectrometers of which the Perkin-Elmer Infracord instruments are so typical.

In 1964—1965 Perkin-Elmer announced that they were developing a laser-powered spectrometer. They were followed quickly by Cary Instruments, Coderg, Spex, and others. It is quite clear that, if so many commercial instrument makers are active in the field, there must be considerable evidence that the laser has effected a complete renaissance. It seems probable that several hundred instruments may well find their way into chemical laboratories during the next few years.

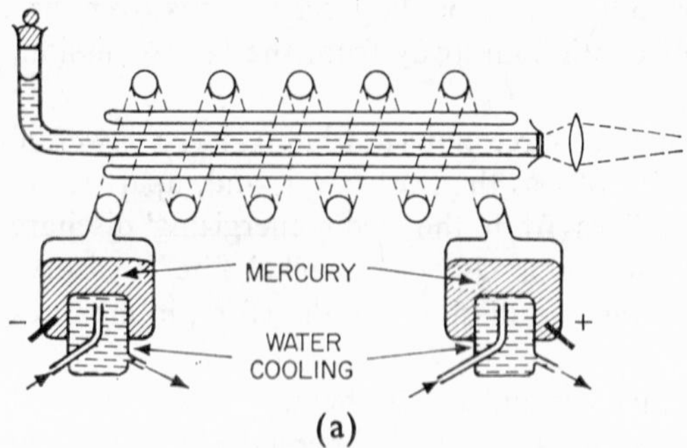

Toronto mercury arc; this would typically consume 25 amp at 100 volts

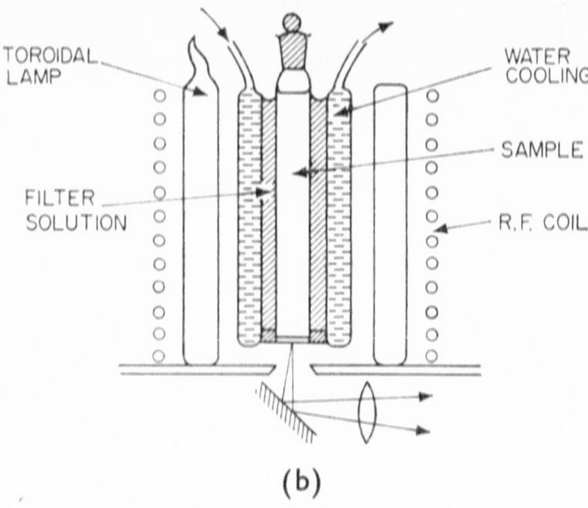

Radiofrequency lamp: this would be powered by ca. 1½ kilowatts of radiation at 5—10 megacycles/sec

Figure 2.1 Two discharge sources used for Raman spectroscopy

2.2 Basic Units of a Modern Raman Spectrometer

2.2.1 *Source*

The primary advantages of the laser sources over the discharge lamp are the lack of continuum away from the laser emission, and the single intense line.

The helium–neon source emits at 6328 Å, has a power of up to 100—200 mW, a line width of less than 0·1 cm^{-1}, and, apart from some relatively weak spurious lines from the neon energizing discharge, virtually no background continuum, Use of a 'spike' filter to isolate the laser line removes this spurious radiation and therefore, in most respects, produces the ideal Raman source.

The ruby source resembles the helium–neon device in these respects, but the ionized argon, krypton, and neon devices are capable of operation at a considerable number of wavelengths. Continuous output can be obtained at the following wavelengths:

Argon	Krypton	Neon
4579 Å	3507 Å	3324 Å
4658	3564	+ others in the visible
4727	4762	
4765	4825	
4880*	5208	
4965	5682	
5017	6471	
5145*	amongst others	

*Much more powerful than the others.

but it is essential to incorporate some 'tuning' device within the optical system of the laser to make it operate at one wavelength only. Not all lasers operate in a continuous mode. The original ruby device can be made to run so that it is pulsed, or pulsed with a high repetition rate (quasi-continuous), or even continuous.[7] In addition, it can be Q-spoiled to operate in a giant-pulse mode. These ideas are elaborated in Appendix 1. If photographic detection of radiation is envisaged, pulsed or quasi-continuous operation fare as well as continuous, but it must be remembered that a Q-spoiled laser emitting 1 megawatt for 10^{-7} sec has power similar to that of a continuous device of only 100 milliwatts exposed for 1 sec. Sometimes the speed of data acquisition resulting from the use of a pulsed laser is an advantage; sometimes it is not. For photoelectric recording of spectra, a continuous (or at least quasi-continuous) device is essential. The helium–neon laser has been popular because its output is very steady ($<\pm 1\%$ variation) as well as being monochromatic. The ionized-gas devices tend to be far less stable, and special construction, particularly of the power supply, is required to obtain acceptable stability ($\pm 5\%$ output power). The beam from a laser may be very highly polarized, especially if the unit is of the gas-discharge type and the discharge tube is closed by two Brewster-angle windows. This property can be useful particularly for the determination of depolarization ratios or the examination of single-crystal spectra.

Another noteworthy point concerns the positioning of the sample. Since the laser is, in effect, an optical oscillator the *output* power is only a small proportion (1–20%) of the power transient within the device. One method of increasing the illumination of a sample would therefore seem to be to place it *inside* the cavity. This has, in fact, been successfully used as a method of excitation but it does have a number of disadvantages: it is not possible to use a 'spike' filter and the sample must be extremely transparent and non-turbid if the laser is not to be extinguished. As a general method it is therefore limited and unimportant.

Another possible source is the neodymium laser followed by a second harmonic generator. This device changes the laser wavelength to half its 'normal' value; in this case 1·046 μ. Emission of ~1 W at 5300 Å has been reported using the generating crystal inside the Nd^{3+} (in yttrium–aluminium garnet) laser cavity. The neodymium emission at 1·3μ could be used to produce continuous radiation at ~6500 Å.

Another possible Raman source is the 'tunable' laser. When a continuous device with a linewidth of ~1 cm^{-1} operating in the range 6000–10,000 Å becomes available (and this will be in a very short time) a revolutionary Raman system will be developed. Part of the radiation from the source will be reflected to a reference detector and the remainder will irradiate the sample. Between the sample and the Raman spectrum detector there will be a narrow-pass interference filter and/or monochromator set at a fixed wavelength. The signal from the main and reference detectors will be compared and the results plotted against wavelength as the source is tuned over the visible and near-infrared.

Discharges emit in all directions, and although numerous attempts have been made to improve their directional properties (e.g. by forcing the discharge through a capillary and 'viewing' along the capillary axis), most discharge lamps have been constructed in an annular or coiled form, the sample being placed at the centre of the lamp. In order to obtain acceptably efficient transfer of the radiation from the lamp to the source, a relatively large sample volume was desirable (frequently several ml, although special microsampling devices enabled this to be reduced to 20 μl in favourable cases). Even at best, however, the coupling was very poor indeed. The laser on the other hand produces its radiation in a well collimated beam of very small cross-section. As a result the flux density (in watts/cm^2) in a small sample can be extremely high. Use of a lens to increase this is highly satisfactory because the beam is almost coherent and therefore the focused beam can be minute. Intensities of 200—1000 W/cm^2 of radiation at the sample have been claimed in some cases when focusing lenses have been used. Obviously it is possible to illuminate a very tiny sample efficiently using these sources but it may not be so easy to collect and analyze the scattered radiation. It is essential to use a very carefully designed optical coupling system between the sample and the slit of the monochromator.

Attempts have been made to compare quantitatively the value of the helium–neon laser and the mercury lamp as Raman sources (ignoring fluorescence and absorption of the existing radiation). Evans, Hard, and Murphy[8] compared a 2 kW mercury lamp at 4358 and 5461 Å with a 30 mW laser, and concluded that there was little to choose between them at shifts of $\Delta\nu = 1000$ cm^{-1}. Detector sensitivity failure at long wavelengths led them to conclude that the blue source is superior to the red at large shifts. Hawes *et al.*[9] have shown that the $\Delta\nu = 459$ cm^{-1} line of carbon tetrachloride has similar intensity from 5 ml illuminated with Hg 4358 Å light from the Cary Toronto Arc (18 amp at 150 V d.c.) and from 0·03 ml in a capillary cell irradiated with ~65 mW at 6328 Å, using the appropriate Cary transfer optics. They also point out and discuss that, whereas the intensity $\propto \nu_0^4$ light-scattering law is widely quoted, the signal/noise ratio in a spectrometer is related to $\nu_0^{3/2}$ thus explaining this similarity in performance. Results obtained at Southampton tend to confirm these observations on clear colourless samples, but the use of the argon ion laser can effect an enormous improvement. This latter device operating at 4880 Å (~500 mW) frequently gives a signal/noise ratio improvement of more than 50 over the helium–neon laser.

2.2.2 *Pre-slit Optics and Sampling Systems*

Almost every laser-Raman instrument seems to have its own unique sampling system and pre-slit optics. However, the basic requirements are the same in each case: to illuminate the sample in such a way that an acceptable image is produced to be passed efficiently to the entrance slit of the monochromator.

Three basic geometries are used:

1. The conventional 90° system where the laser beam passes normally to the direction of view.

2. The 180° system where the beam is travelling away from the slit. When highly reflecting materials are being studied, an appreciable proportion of the laser radiation will pass into the monochromator, so a very elaborate spectrometer will be essential.

3. Seemingly the most disadvantageous, the 0° system where the laser beam is directed at the slit. Although it may be possible to deflect the beam away from the monochromator by use of a small mirror before the slit, Tyndall scattering will tend to make this reflector ineffective.

Surprisingly, all three systems have been used with success. A few acceptable arrangements are shown in Figure 2.2, and these will be discussed in turn.

(a) The laser is focused into a capillary or other tube containing a liquid or polymer sample. In principle a sufficiently perfect tube may also lie *across* the direction of the laser beam. The source object is slit-shaped, and therefore it is easy to focus the scattered radiation on to the entrance slit of the monochromator.

(b) A system devised by Perkin-Elmer for their LR-1 spectrometer. The beam is reflected up and down the cell between two inclined reflectors. In favourable cases 150 passes can be achieved before the beam dies away owing to absorption. Again the image is slit-shaped.

(c) A system used by Spex Industries for studying solids as pellets. In white solids the laser is largely reflected at the monochromator. The object is elliptical and small, thus efficient transfer of the radiation to the monochromator is poor. It is sometimes convenient to hold a powder sample in a glass vessel and admit the (axially polarized) laser at approximately the Brewster angle.

(d) A Perkin-Elmer development for studying solids. If the sample is very dark in colour the laser radiation is absorbed so much that no spectrum is obtained.

(e) The Cary 81 system. The liquid is contained in a small-bore capillary. The scattered radiation is passed to the collector lens and on to the monochromator by multiple reflections inside the capillary. Unfortunately the polarization of the Raman light is partially destroyed, so depolarization data recorded may be difficult to interpret. The object shape is round but the Cary instrument incorporates an image slicer to compensate for this property.

(f) The solid-phase equivalent of (e). However, no valuable depolarization data can be obtained for powders or opalescent bulk solids.

(g) A hypothetical arrangement which is related to the Cary system.

(h) A 0° system where the beam is reflected back towards the laser. The scattered radiation is collected and passed to the monochromator.

(i) An alternative to (h) for small samples. The method is poor if the sample is absorbing and the intensity of the source line relative to the Raman line is very high.

Of all these systems, the only ones which do not require the sample to exist in a special form (pellet, liquid in a capillary, etc.) are (f) and (g). In arrangements (c) and (f) the beam is not required to pass through the specimen, and so they are particularly valuable for the study of opaque solids.

In some of these arrangements [(c)—(i)] the image is very small and round. A device which collects a moderately large amount of the available light does so over a large solid angle. This is possible if the source object is small. For example, in the Cary spectrometer radiation is collected over a solid angle of 66°. The difficulty with this somewhat idealized principle is that the image of the scattering sample is round. The shape of the image can be changed by use of an image slicer. The Cary 81 is the only commercial instrument which incorporates this device. The image is split into twenty vertical strips which are then re-oriented so that they are placed one above the other, thus converting a round image into a long slit-shaped image. Since the instrument uses a 4 inch high entrance

Figure 2.2 Some of the sample arrangements found in laser Raman spectrometers; L signifies the incoming laser beam, and the broken arrows indicate the rays proceeding to the monochromator. Note that in cases (b), (d), and (h) the laser need not be focused on the sample; in (e) and (f) a convergent beam is used

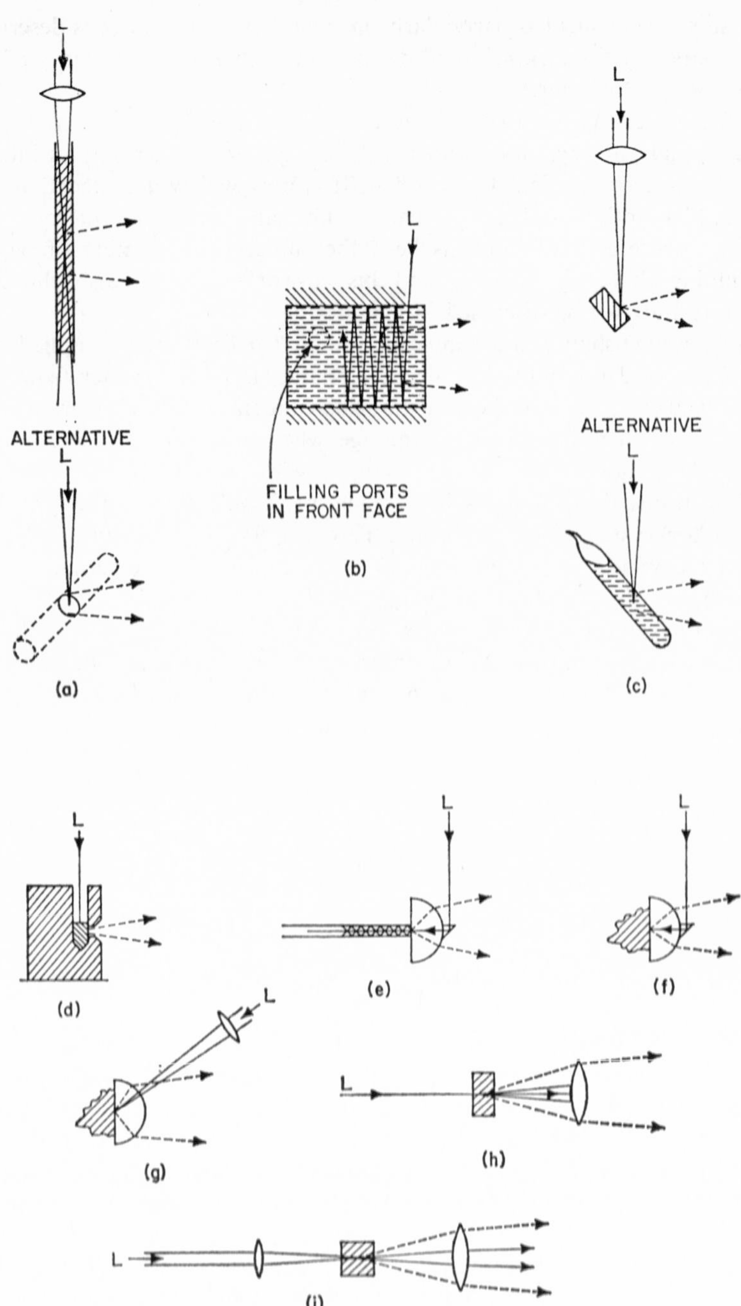
L
ALTERNATIVE
L
(a)
L
FILLING PORTS
IN FRONT FACE
(b)
L
ALTERNATIVE
L
(c)
L
(d)
L
(e)
L
(f)
L
(g)
L
(h)
L
(i)

slit this arrangement is particularly efficient. The Cary device is described in a paper by Cary *et al.*[10] and discussed together with the theory of the Raman spectrometer.

If the laser beam is focused into a minute sample it is obvious that the radiant flux through the sample will be high, and as a result an intense Raman spectrum might be expected. It is now widely accepted that this is so. The optical problem of beam condensation and efficient collection of the scattered radiation has been the subject of a number of papers including those by Delhaye and his co-workers[11] and also those of Barrett and Adams discussed in Chapter 5.

The wedge-shaped interference layer coated cells often supplied and recommended for multiple reflection of the laser beam suffer from two disadvantages: the laser beam cannot be condensed, and they are neither robust nor disposable (a disadvantage when used for corrosive or air-unstable samples).

For transparent samples it is much preferable to use the 'external resonator' system where the beam is reflected between a separate dielectric mirror beyond the sample and the laser mirror. The external mirror may be concave to allow for the presence of a focusing lens, and in this way all passes are focused at the same point. An additional plane reflector can also be inserted just off the primary beam, to 'catch' the reflected beam. Up to 7-fold enhancement appears to be obtainable with no cell in position (i.e. for an air spectrum), but manufacturers' literature does not adequately stress the importance of non-reflecting cells and an uncluttered laser path (or an additional mirror). However, two- or three-fold enhancement may be obtained with ordinary Pyrex tubing cells, which have the advantage of being both simple and disposable.

The 'lining-up' of this system can be very tedious.

2.2.3 *Monochromators*

It has been pointed out already that the Raman bands are extremely weak relative to the radiation at the exciting frequency ν, and these bands are, of course, close to ν in frequency. If one is to study opalescent materials, or to use optical arrangements other than the ideal, the ratio I_ν/I_{Raman} becomes even more unfavourable. It is therefore essential to use a monochromator system with very high discrimination. If a monochromator is illuminated with a perfect monochromatic source of frequency ν, the radiation leaving the system should theoretically be zero until the monochromator is scanned across ν, when the signal will maximize and then return immediately to zero beyond ν. In practice a *weak* signal will be detected *away* from ν, since stray light will always be present inside the instrument. The ratio $I_{\Delta\nu}/I_{\nu}$ is used as a measure of the 'discrimination' of the monochromator. Typically, at $(\nu \pm 50)$ cm^{-1} a value of 10^{-5} is accepted as a reasonable value for this ratio. It is said that glass-prism dispersion monochromators may perform slightly better than this. Since, in the most favourable example, a Raman line will rarely have more than 10^{-2} of the intensity of the Rayleigh line or 10^{-6} of that of the source, this performance leaves little in reserve. In 1965 Landon and Porto[12] suggested the use of a tandem spectrometer, i.e. two identical spectrometers in series, to improve on this situation. They showed that a stray-light discrimination of $\sim 10^{-10}$ could be achieved by this means.* However, it is noteworthy that an instrument using a double monochromator (the Cary 81) had been commercially available since 1961.

The efficiency of a monochromator, as far as its energy transmission is concerned, is a function (amongst other things) of the slit height and the focal length.[15] It is therefore desirable to use as high a slit as possible. Most of the instruments described in the literature and available commercially have slits of between 0·5 and 1·0 inch since longer slits could not be efficiently illuminated. The Cary spectrometer uses a uniquely large slit, 4 inches high, and this in turn necessitates the use of the image-slicer already described.

By far the most popular optical design is the Czerny–Turner device described in detail by Fastie.[15] This system has the advantage that concave spherical mirrors are used, and these are relatively inexpensive. Two commercial machines, the Cary 81 and the Perkin-Elmer LR-3, use the Littrow system which is more familiar in infrared spectrometers. In these, off-axis ellipsoidal reflectors are used to collimate the beam. It is not clear at present which of the two, the Littrow or the Czerny–Turner system, is the more efficient as a Raman monochromator.

---

*Messrs. Jarrell-Ash have given the following figures for their Czerny–Turner double monochromator system:[13]

Distance ($\Delta\nu$) from parent line	~ 12	~ 25	~ 100	cm^{-1}
$I_{\Delta\nu}/I_{\nu}$	$2{\cdot}2 \times 10^{-9}$	$8{\cdot}0 \times 10^{-11}$	$<1 \times 10^{-11}$	

Spex Industries have produced the following data for their Czerny–Turner model 1400 double monochromator:[14]

Distance ($\Delta\nu$) from parent line	~ 12	~ 25	~ 100	>150	cm^{-1}
$I_{\Delta\nu}/I_{\nu}$	5×10^{-8}	5×10^{-10}	$<1 \times 10^{-12}$	$<1 \times 10^{-14}$	

2.2.4 *Detectors, Amplifiers, and Recorders*

Photoelectric recording is achieved normally by the use of a photomultiplier, amplifier, and recorder. In a few of the 'home-built' instruments, photographic detection is still used for simplicity but, although this system is cheap, the time necessary to develop plates and examine them with a microphotometer, to produce a curve of emission against $\Delta\nu$, makes the method unsuitable for routine spectroscopy.

Photomultipliers for laser Raman work are always of the low-noise type and are sensitive at longer wavelengths. A photocathode of the S1 or (for red lasers) the S20 type is normally used.* These tubes, although sensitive out to 8500 Å, show a very steep fall-off in performance from the green to longer wavelengths. As a result, with the helium–neon laser, performance of the spectrometer falls off rapidly to the Stokes side of the exciting line. From experience with the Cary 81 spectrometer at Southampton, this fall-off is at least by a factor of 0·5 at $\Delta\nu = 1000\ cm^{-1}$ and could be as much as 0·02 at $\Delta\nu = 3200\ cm^{-1}$. It is probable that this property of the Cary is shown by other instruments. If it were intended to use the quasi-continuous ruby source, good spectra could be recorded only close to the exciting line before detector performance would attenuate the results unacceptably. Unfortunately, in general, infrared detectors, of which the lead sulphide photoconductor is typical, have not sufficiently good signal/noise ratios to make them useful for Raman spectroscopy. Silicon semiconductor devices are also useful for the near-infrared (the typical range is from the visible to $1{\cdot}2\mu$, with maximum sensitivity at ~9000Å). They are small and at present the noise is high, but improvements occur very rapidly. Using the ruby laser line, good performance can be achieved since, even at $\Delta\nu = 3500\ cm^{-1}$ (Stokes), the detector will have maximum sensitivity.

Loader has compared the argon laser with the helium–neon laser in producing the spectrum of chloroform in a sealed capillary cell on a Cary 81 instrument:

	$\Delta\nu\ (cm^{-1})$	He–Ne laser	Ar^+ laser	Improvement
ν_3	366	63	730	× 12
ν_1	3019	10	5970	× 597

Spectral slit-width 5·0 cm^{-1}. Intensity units arbitrary.

---

*Tubes equivalent to R.C.A.-7102, E.M.I.-9558A, and I.T.T.-FW130 are popular for He–Ne laser sources. To obtain the best performance the tubes are always especially selected for red sensitivity. New phototubes involving semiconductor dynodes have been announced recently by R.C.A. These appear to have excellent red sensitivity.

and also the spectrum of n-octylphenol solid powder:

He–Ne laser	Gain of 500 arbitrary units to obtain a spectrum out to $\Delta\nu = 1500\ cm^{-1}$.
Ar^+	Gain of 100 units required. Noise so low that very fast scanning was possible.

Some further comparisons supplied by Ozin[16] may be of value:

Niobium pentoxide (Nb_2O_5) powder	× 50 improvement using 4880 Å source.
Saturated solution of sodium dithionite	× 100 improvement relative to the He–Ne device.

Blue or green compounds give very poor spectra (if at all) when red sources are used, but the Ar^+ laser can be most satisfactory. Excellent spectra have been recorded for the $CoCl_4^{2-}$ ion in Cs_3CoCl_5 (deep blue), the compound $CuCl_2,2H_2O$, and vanadyl acetylacetonate (very deep blue).

In the above experiments the He–Ne laser was a Spectra Physics type 125 (~65 mW) and the Ar^+ was a Spectra Physics type 140 instrument tuned to 4880 Å with power ~1 W or possibly a little more.

It has been reported (and in some places denied!) that cooling of detectors can be very beneficial in enhancing their signal/noise ratios, particularly at red wavelengths. By using radiation at 8521 Å (from a Cs hollow-cathode lamp) engineers at the Jarrell-Ash Co. have demonstrated the following improvements in performance for an R.C.A.-7102 tube incorporating their thermoelectric photomultiplier cooler assembly:

Temperature	ambient	0	−10	−25	°C
Signal	0·4	0·04	0·04	0·04	$\times 10^{-6}$ amp
Dark current	7000	220	30	7	$\times 10^{-9}$ amp
Signal/noise	0·06	1·5	13	57	

Other sources of information on this aspect are given in ref. 17. The efficiency of cooling in reducing noise or increasing signal varies greatly from one photomultiplier to another. However, it seems that two effects always occur on cooling: the noise is reduced and the sensitivity of the photocathode falls. Thus the signal/noise ratio may not improve with cooling. Nakamura and Schwarz demonstrate this point nicely in a paper on photomultipliers and signal processing systems.[18] Experience at

Southampton tends to confirm that the signal/noise ratio is not very sensitive to temperature. It is probably significant that Coderg and Spex have joined Jarrell-Ash in offering a detector cooler for their Raman spectrometer.

A number of amplification systems are in current use. In the most popular, lock-in or synchronous amplification, the optical beam is interrupted before the detector at a low audiofrequency, and the a.c. signal from the detector is amplified by a phase-sensitive amplifier, filtered to remove 'noise', and rectified to produce an output d.c. potential proportional to the intensity of the radiation to be detected. Although the system operates satisfactorily, the more elaborate 'photon counting' technique is rapidly finding favour in Raman instruments.

Pulses from a photomultiplier due to noise are primarily of high and low energy, whereas the pulses due to the signal tend to have energies between certain limits. In the photon-counting technique an energy sorter or 'gate' is inserted after the detector to 'pick out' the pulses due to the optical signal, and it is followed by a system to attenuate the pulse energies to a constant value and a rectifier to produce a d.c. voltage proportional to the number of photons per second arriving at the photomultiplier. Another system, that of 'intensity correlation', has been described but has not been applied to laser Raman spectroscopy.[19]

In some instruments, an even simpler system is used, that of d.c. amplification followed by filtering and recording.

Synchronous detection, d.c. amplification, and photon counting have been compared by Nakamura and Schwarz.[18] At low light levels (below the dark-current equivalent value) the pulse technique is certainly the most satisfactory, but at higher intensities there is little to choose between the three detector–amplifier systems. It is the opinion of many who use both systems that, at more realistic light levels, the synchronous detection system has much in its favour since in photon-counting there is a tendency to pick up electrical interference from outside.

Recording presents little problem, since an enormous range of millivolt recorders can be bought. However, for routine spectroscopy it is not adequate to use a strip-chart recorder and indicate the frequency ($\Delta\nu$) values by printing on to the chart a pulse fed from the wavelength drive. In Raman spectroscopy it is essential to be able to examine bands again and again in order to choose the optimum slit and gain settings, and therefore the chart must be mechanically synchronized with the wavelength drive. This feature has always been incorporated on the Cary spectrometer.

2.3 Historical Development of Laser Raman Spectrometers

The first experiments in the field were completed by Porto and Wood[20] and by Stoicheff[21] during 1962. The first report described how a pulsed ruby laser was focused into a glass cell containing carbon tetrachloride or benzene. The cell was coated with barium sulphate to improve optical efficiency. The scattered light was collected by a lens and passed to a low-dispersion grating spectrograph loaded with type 1N film. About fifty pulses were required to complete the experiment. Stoicheff used a multiple-reflection cell and a spectrograph to examine carbon disulphide. The ruby laser is not a very convenient source because a number of pulses are required and yet the device can be pulsed at most only a few times per minute, but its long wavelength can be an advantage especially for coloured systems.

In the following year Daniliheva *et al.*[22] demonstrated not only the potential of the ruby laser as a source but that we now have available a powerful source for studying deeply coloured materials. They studied 4,4′-azoxyanisole and anisal-*p*-aminoazobenzene successfully using 100 flashes from a ruby at 1—1·8 Joules and with photographic detection.

It was clear that the use of a continuous laser would be a major step forward. By 1963 Kogelnik and Porto[23] had demonstrated the use of the helium–neon device by examining carbon tetrachloride, benzene, and carbon disulphide in a cell *within* the laser cavity. They used photographic detection (with Polaroid infrared-sensitive film) and a simple grating spectrograph. The laser used gave 5 mW output as set up for the experiment ($\sim 200 \times 5$ mW of laser power passed through the sample), and an exposure of 1 hour was required (20 sec was sufficient for the strongest lines in benzene). Clearly the experiments were not applicable to absorbing or turbid samples since these would simply damp-out laser action. Kogelnik and Porto also utilized the fine polarization properties of the gas laser, and pointed out the potential of the device in this respect. Also during 1963 Porto, Cheesman, and Siqueira[24] had continued and developed their interest in the ruby as a source, and had devised and tested numerous sample geometries. Some of these were given in a later review by Porto.[25]

Using a continuous laser, photoelectric recording of Raman spectra is an obvious development. Leite and Porto[26,27] were the first to achieve this when, in 1963 and 1964, they recorded excellent spectra from benzene placed inside the cavity. They were closely followed by Koningstein and Smith[28] and also by engineers at the Perkin-Elmer Corporation[29] who developed their prototype LR-1 commercial spectrometer in 1964. These two sets of experiments involved placing the sample external to the laser cavity, and they are therefore much more versatile in application. Scattered light in both cases was analyzed by single-grating spectrometers and photomultipliers. Also during 1964 some British physicists at the National Physical Laboratory, Teddington, demonstrated the use of the interferometer in Raman spectroscopy.[30] They illuminated their samples with a gallium arsenide injection laser (at 0·84 μ) and examined the scattered light with a Fourier-transform interferometer system

By 1965 Weber and Porto[31] had amply demonstrated the value of the helium–neon laser as a source in high-resolution work by giving a spectrum of methylacetylene recorded photographically in 58 hours at a dispersion of 2 cm^{-1}/mm on a sample at 0·5 atm pressure. Also during 1965 the ruby laser was increasingly applied to the study of small specimens of coloured solids. In 1965 and 1966 Schrader and Stockburger[32,33] described their apparatus, surveyed their experiments, and gave a series of spectra of coloured organic molecules. Meanwhile in France, Delhaye and Migeon were similarly active using ruby lasers and photographic plate detection.[11,34]

Since 1965 it has become clear that a considerable reduction in sample size is possible if laser sources are used. The prototype Cary 81 laser spectrometer was fully developed during 1965 and 1966, and this enabled liquid samples of only 0·02 ml to be studied efficiently in glass capillaries. In 1966 Lau and Hertz[35] developed liquid cells of $<$0·2 ml capacity for use with laser sources and spectrographs, and they also described the use of cells in 'external resonator' cavities of a helium–neon laser.

The quasi-continuous ruby laser was first used as a source during 1966,[36] whilst experiments with the argon ion laser appeared during the same year.[37] As forecast, Porto and his co-workers used the excellent directional and polarization properties of the laser which were put to use as early as 1964 when Leite, Moore, and Porto obtained accurate depolarization ratios on carbon disulphide, benzene, toluene, and other molecules.[38] In the following year Damen, Leite, and Porto studied the angular dependence of the scattering from benzene using a helium–neon laser.[39]

2.4 Commercially Available Spectrometers

A very brief specification is given for each of the commercially available laser Raman spectrometers known to the authors. They are listed alphabetically.

2.4.1 *Cary 81* Cary Instruments, Monrovia, California, U.S.A.

Source. Spectra Physics type 125 He–Ne laser, ~65 mW; Spectra Physics Ar^+ laser available as an optional extra.*

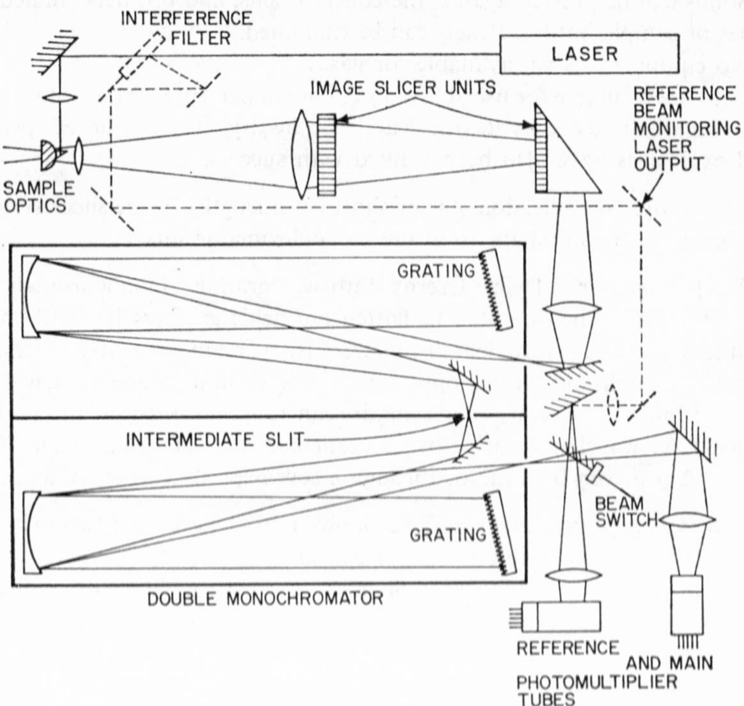

Figure 2.3 The Cary 81 spectrometer (elevation)

Sample system. 180 degree system normal; facilities for 90° viewing incorporated; collection over a wide angle (~66°); depolarization measurements made using a half-wave plate before the sample and a Polaroid analyzer after it.

Liquids should be in capillaries for best results; glass or quartz are

*In common with other instrument makers, Cary seem willing to have other lasers fitted to their instruments. Two which must be mentioned because of their popularity are the Coherent Radiation Ar^+ and Kr^+ lasers and the Carson Ar–Kr device. Details of the latter can be obtained from Spex Industries, and of the Coherent Radiation Laboratories devices from them at 932 East Meadow Drive, Palo Alto, California, U.S.A.

suitable and they can be connected permanently to a gas line. If they are good scatterers (pure liquids etc.) samples can be in ampoules or bottles and viewed through the glass. A multireflection cell (2·5 ml capacity) is available for 90° viewing.

Solids can be pressed against the collector lens, and powders studied in tubes or sample bottles. Pellets can be examined.

No equipment is yet available for gases.

Cells are available for use at low and high temperatures. Many 'unusual' samples are discussed by Beattie[40] and also by Bryant.[41] Adsorbed species and explosives have also been studied with success.

Pre-slit optics. Collection lens of short focal length; image slicer system to change the image shape to fit the monochromator slits.

Monochromator. Twin Czerny–Littrow grating monochromators; corrector lenses incorporated to flatten the field; gratings 100×100 mm with 1200 lines/mm; focal length 1 metre; dispersion in red ~10 cm^{-1}/mm; slits 10 cm high and curved optionally single or double at the turn of a switch (Shurcliff system); wavelength scan range throughout the visible to 8500 Å; wavelength drive by cosecant bar (Plate I, facing page 50)* (linear $\Delta\nu$) driven by a motor through a seven-speed gearbox system.

Detector–amplifier–recorder. Two photomultipliers for the two exit slit images plus one for a reference optical train from the laser; signal from the latter used to attenuate main signal to correct for fluctuations in the source power; a.c. null-balance system used to drive a recorder mechanically linked to the monochromator scan drive.

Price. ~43,000 U.S. dollars (~£16,500).

Construction of capillary cells for the Cary 81. Cary Instruments Corporation state that these cells should be constructed of silica to reduce fluorescence, and that a good flat end is necessary for best results. We note that it is the Raman rather than the fluorescence spectrum of glass which causes trouble. This is proved immediately by the fact that the same shape of background is observed with any sufficiently intense laser line as exciting source. The background is also largely polarized, which would not be expected for fluorescence from a glass.

It has been found at Southampton that the interference caused by the Raman spectrum of glass (particularly by a sharp drop in emission at

---

*The grating formula is $n\lambda = 2d \sin\theta$, where n = integer, λ = wavelength, d = grating line spacing, θ = scan line. To get a linear cm^{-1} scan it is necessary to change the angle θ linearly in cosec θ.

about $\Delta\nu = 500\ cm^{-1}$) can be reduced to an acceptable level for a Pyrex cell if a very thin window is used. Furthermore, this window can easily be obtained by blowing and grinding, as we shall describe. By this method cells have been constructed which slightly surpass the 20-dollar Cary product in light-gathering power (since they are longer) although the depolarization correction is also greater. Spills of Pyrex glass tubing of appropriate diameter can be purchased, or drawn in the usual way. Referring to Figure 2.4 (a) the tubing is first drawn to a *symmetrical* point in a fine flame and the end is knocked off. (b) It is then put just into the side of a non-turbulent flame whilst being rotated until the effect shown in (c) is obtained. The shape to aim for is a flat *inside* surface, and this is

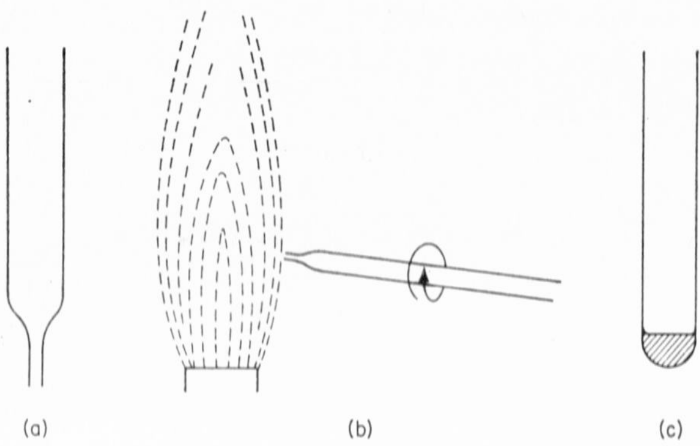

Figure 2.4 Method for making capillary sample tubes for the Cary 81 spectrometer

easily achieved with practice. Finally the end may be ground off using fine (600 grade) carborundum powder and water on a glass plate in the usual way. It has been found that it is not absolutely necessary to polish this fine-ground surface, provided that it is put into optical contact with the hemispherical lens using glycerol.

2.4.2 *Coderg PH1* Société de Conversions des Énérgies, Clichy, France

Source. Uses commercially available lasers of which the following are recommended: Ar^+ 4880 Å 200 mW, manufactured by C.S.F. (France); He–Ne 6328 Å 50 mW, manufactured by C.G.E. or O.I.P. (France); ruby

6943 Å 100 mW, manufactured by Siemens. Also available with Spectra Physics type 125 (He–Ne 65 mW) or 141 (Ar^+ 100 mW at 5145 or 4880 Å).

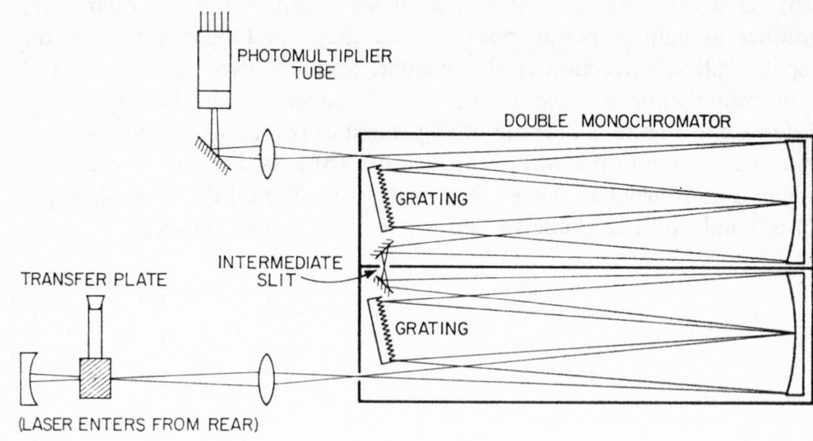

(a) Elevation

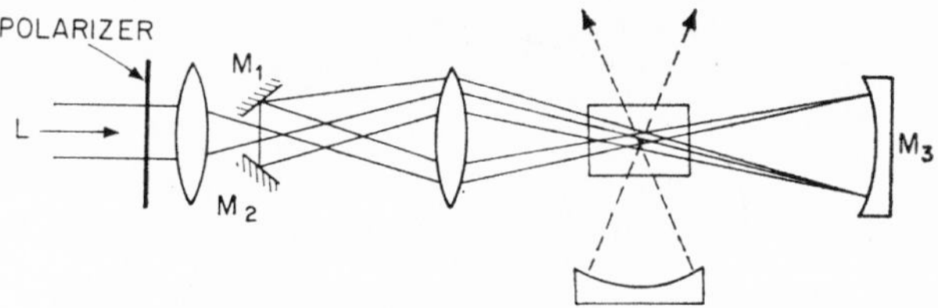

(b) Alternative transfer plate, for liquids and powders, involving a multipass system with mirrors M_1, M_2, and M_3

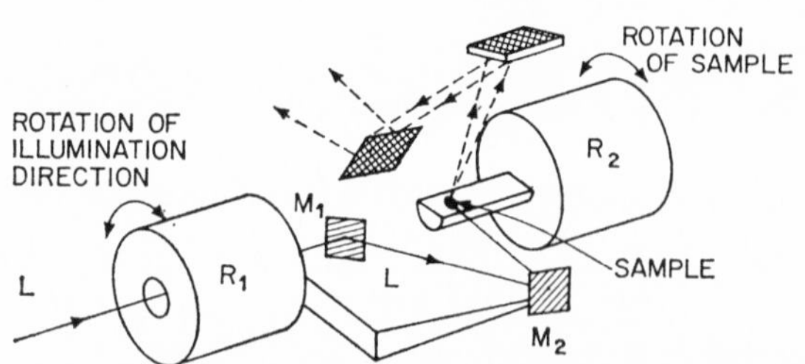

(c) Alternative transfer plate in which the mirrors M_1 and M_2 are carried by the rotor R_1, and R_2 carries the sample

Figure 2.5 The Coderg PH1 instrument

Sample system. A series of transfer plates are used involving 90° viewing for examining liquids, solids, and single crystals. Assemblies for obtaining spectra at high pressures and low temperatures are also available.

Equipment available for gases.

A number of simply constructed cells for studying deeply coloured species such as powders have been developed by Clase.[42] Details of many of these are to be found in a book by Loader.[43]

Monochromator. Double Czerny–Turner design; aperture *f*/8; focal length 600 mm; grating ruled with 1200 or 1800 lines/mm; resolution of 0·3 cm^{-1} is claimed; scan is linear in cm^{-1} by cosecant bar; specially designed slits (opening stepwise) incorporated; scan speeds (stepwise) 1—2000 cm^{-1}/min; range 4000—8500 Å.

Detector–amplifier–recorder. Photomultiplier with S20 photocathode used; amplification system by 'continuous' (d.c.), 'noise voltage', or photon-counting systems; recorder type RE 511 (commercially available); wavenumber indication printed on to the chart; recorder not coupled to the wavelength scan system.

Price. ~180,000 francs (~£14,000) complete with either He–Ne laser or Spectra Physics type 141.

2.4.3 Raman "*Automatique*"

A convenient small table-top instrument.

Laser. Spectra Physics type 124 (15 mW) He–Ne built into the machine.

Sample systems. Similar to the PH1.

Monochromator. Double Czerny–Turner design; aperture *f*/6; focal length 300 mm; grating ruled with 1200 lines/mm; resolution of 1 cm^{-1} claimed; linear cm^{-1} scan.

Detector–amplifier–recorder. Photomultiplier S20 type; amplification system by d.c. or 'noise voltage' systems; recorder as for PH1, not coupled to $\Delta\nu$ scan.

Price. ~98,000 francs complete (~£8000).

2.4.4 Raman "*Didactique*"

A further machine of very low price and table-top design.

Laser. He–Ne; power 10 mW.

Monochromator. Single grating; aperture *f*/6; focal length 300 mm; grating ruled with 1200 lines/mm; resolution 4 cm^{-1}; linear cm^{-1} scan.

Detector–amplifier–recorder. Photomultiplier S20; d.c. amplifier; Metrocord recorder.

Price. ~39,000 francs (~£3200) without laser.

2.4.5 *Huet R50* Société Générale d'Optique, Paris 19e, France

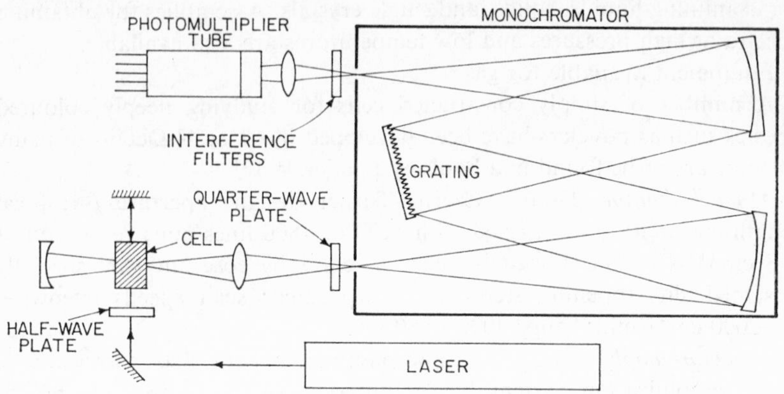

Figure 2.6 The Huet R50 spectrometer (plan)

Source. 50 mW He–Ne laser by C.G.E. of France mounted behind the spectrometer.

Sample systems. Rectangular liquid cell through which laser beam passes twice; viewing at 90°; depolarization using half-wave plate before the sample and quarter-wave plate scrambler before slit; cell sizes 5, 2, 1, 0·5, and 0·2 cc; multipass cell for solutions of low concentration.

Monochromator. Single Czerny–Turner design; aperture *f*/4·5; focal length 450 mm; grating 64 mm^2 with 1200 lines/mm, blazed at 7500 Å; dispersion 17 Å/mm; $\Delta\nu$ scan by cosecant bar; ten speeds from 0·4 to 410 cm^{-1}/min; interference filters used before the detector; chopping frequency 330 c/sec.

Detector–amplifier–recorder. Photomultiplier with ten-stage S20 photocathode; phase-sensitive detector–amplifier system (max. gain 100 db) feeding a 10 mW separate desk recorder; continuous check on the output of the laser indicated automatically.

The complete spectrometer is mounted on a knee-hole desk with the electronic components beneath the desk-top.

Price. Apply to manufacturer.

2.4.6 *Jarrell-Ash 25-300* Jarrell-Ash Co., Division of Fisher Scientific Co., 590 Lincoln St., Waltham, Mass., U.S.A.

Source. He–Ne laser ~60 mW normally fitted but any device of <5 W can be accommodated at will.

Sample system. Very versatile; can illuminate sample from above or below (90° system) or horizontally along the axis of the entrance slit; adjuster from 0 to 9° to this axis also available; sample compartment is very large (64×87×67 cm); half-wave plate and analyzer used for depolarization measurements; wedge-type depolarizer before slit; instrument has been used to study compressed gases, crystals, molten salts, and also high-resolution spectra of gases.

Monochromator. Instrument incorporates a type 25-102 Jarrell-Ash double Czerny–Turner monochromator; monochromator thermostatted; gratings mounted one above the other on a common pivot axis; scan by cosecant bar (linear cm^{-1}) driven by a twelve-speed gearbox and reversible motor at speeds between 1000 and 0·2 cm^{-1}/min; monochromator range 21,900—11,500 cm^{-1}; slit height 20 mm; servo-operated slit system available.

Detector–amplifier–recorder. A cooled I.T.T.-FW130 type S20 photomultiplier; cathode slit-shaped; operating temperature −22°C (cooling gives a 100-fold improvement on the detector's performance at 20°C); photon-counting used to analyze the signal; pre-amplifier built into detector socket; conventional strip-chart recorder. A 'blip' system is used to indicate $\Delta\nu$ values on the recorder chart, i.e. the recorder is not coupled to $\Delta\nu$ scan.

Price. 44,900 U.S. dollars. Slit servo-system 3000 dollars extra.

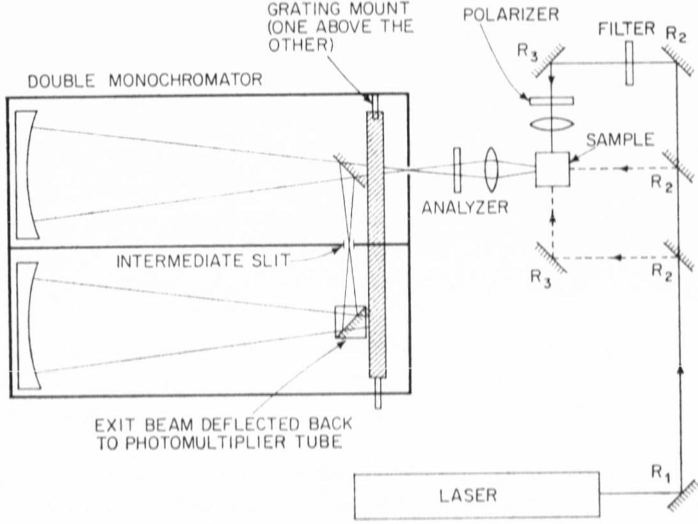

Figure 2.7 The Jarrell-Ash 25-300 laser Raman system (elevation); reflections at R_1, R_2, or R_3 are by Brewster angle prisms; the broken lines represent alternative illuminating directions

2.4.7 *JRS-01A*

Japanese Electron Optical Company,
3-2 Marunouchi, Chiyoda-ku, Tokyo, Japan

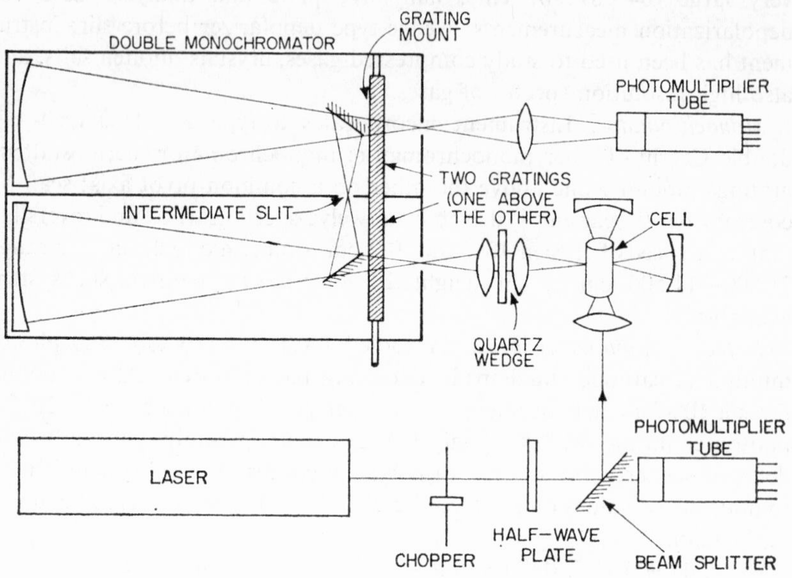

Figure 2.8 The JRS-01A spectrometer (elevation)

Source. Argon ion laser normally built into the machine; output 250 mW at 4880 or 5145 Å; length of laser 1 m. Helium–neon laser available on request; length 2 m; output 50 mW at 6328 Å.

Sample system. Illumination from below; light collected by mirrors and lens; 90° system is normal; polarization measurements using a half-wave plate before and a Grand–Thompson prism after the sample; polarization scrambler fitted before the slit; special low-temperature cells available as well as conventional cells for liquids, powders, and single crystals.

Monochromator. Double Czerny–Turner design with two gratings one above the other; focal length 750 mm; grating 110×110 mm with 1200 lines/mm, blazed at 5000 Å; scan range 2000—8000 Å (N.B. this does not permit operation out to 3500 cm^{-1} from 6328 Å); resolution of monochromator ~0·2 Å; slits 20 mm high; scan by sine drive (linear wavelength) and a marker is fed to the recorder every 5 Å; wavelength indicated to five figures on a digital counter; variable scan speeds available; camera fitted which accepts quarter-plates, 35 mm film, or Polaroid.

Detector–amplifier–recorder. An HTV R292 photomultiplier detector is used, and photon counting and amplification are incorporated, with compensation for laser instability. A 10 mV conventional chart recorder completes the chain.

2.4.8 *Perkin-Elmer LR-1* Perkin-Elmer Corp. Norwalk, Conn.,U.S.A.

In many respects the LR-1 is the "grandfather" of laser Raman machines. It first appeared in 1964 and a considerable number were supplied during 1965—1968. It has been superseded by the LR-3.

Source. Perkin-Elmer He–Ne laser; 6 mW power.

Sample system. 90° system used; collection system simple; depolarization using an Ahrens prism and a scrambler between the source and slit.

Liquids are held in a multireflection cell of volume 2·5 ml. Microcell available (0·2 ml) again using multireflection. A small-volume liquid cell has just been announced; the capacity is 25 μl and the performance ~500% that of the large-volume multireflection cell, i.e. it tends to supersede the 0·2 ml cell; the cell is of quartz and, like the 2·5 ml device, is closed by Teflon plugs.

Solids as powders are pressed into a stainless-steel cylinder. Crystals are easily viewed. A capillary cell for studying solids and liquids has also been developed by Bailey, Kint, and Scherer.[44] The smaller the capillary the less accurate the depolarization ratio.

No cells for gases have yet been announced.

Monochromator. Double-pass Littrow mounted grating (1440 lines/mm, blazed at 6200 Å); 13 c/sec chopping between passes (Walsh system); reflectors interference layer coated; various scan speeds between 4 and 440 cm^{-1}/min.

Detector–amplifier–recorder. S20 photomultiplier; a.c. amplification; strip-chart recorder not coupled to the wavelength scan system; pulse calibration system to recorder.

Price. ~23,000 U.S. dollars (~£10,000).

N.B. Perkin-Elmer have shown spectra recorded using their E.1 monochromator with the sample illuminated by a Spectra Physics laser, but this ensemble does not appear to be available as a complete laser Raman system.

Professor E. B. Bradley has made a number of highly successful modifications to an LR-1 spectrometer, including the use of a different scanning motor (enabling one to scan in both forward and reverse directions), a new pre-amplifier which improves the signal/noise ratio, a cooled photomultiplier, new interference filters, and a Jarrell-Ash quarter-metre Ebert monochromator as a 'pre-selector'. For full details, Professor Bradley should be approached at the University of Kentucky, Lexington, Kentucky 40506, U.S.A.

2.4.9 *Perkin-Elmer LR-3* (Plate II, facing page 51)

Source. Normally a small Perkin-Elmer He–Ne laser but larger ones can be fitted; the instrument will operate from an Ar^+ laser if suitably coated mirrors are fitted throughout the spectrometer.

Sample system. As for model LR-1; an analyzer prism and polarization scrambler fitted in front of the monochromator slit.

Monochromator. Double-pass Littrow mounted grating (1440 lines/mm, blazed at 6200 Å) with 13 c/sec chopping between passes (Walsh system) followed by another grating Littrow system scanned in tandem; reflectors interference coated; various scan speeds between 4 and 440 cm^{-1}/min.

Detector–amplifier–recorder. End-view S20 photomultiplier; a.c. amplification; strip-chart recorder, not coupled to wavelength scan mechanically; read-out on to printed charts identical in scale to those used on the Perkin-Elmer 521 infrared machine; as a result both spectra can be recorded on the same chart.

Price. ~25,000 U.S. dollars (~£10,000). Polarizer, analyzer, and scrambler are extra (1200 dollars).

Figure 2.9 was kindly furnished by Perkin-Elmer. An experimental slit programmer was in use on the LR-3 when these data were recorded; this removes a low-frequency 'shoulder' on the exciting-line band.

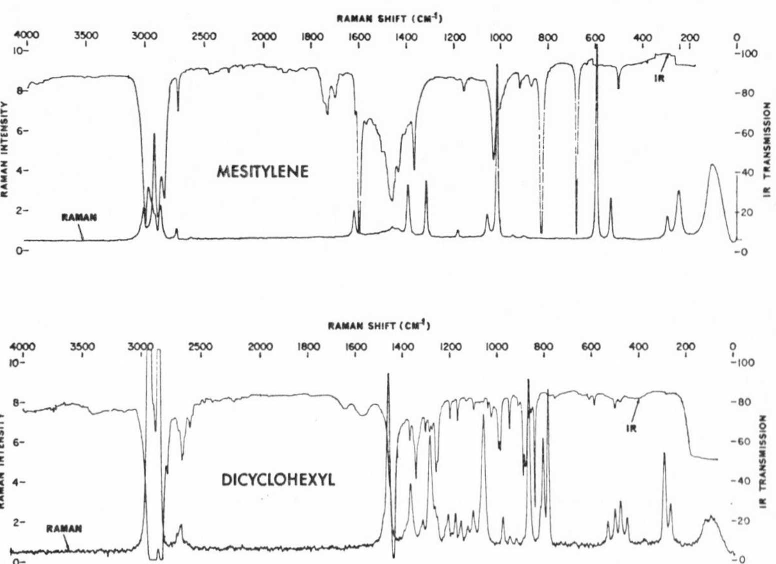

Figure 2.9 Infrared and Raman spectra recorded on the same chart, using Perkin-Elmer 521 (infrared) and Perkin-Elmer LR-3 (Raman) instruments

2.4.10 *Spectra Physics 700*

Spectra Physics,
1250 West Middlefield Road,
Mountain View, California, U.S.A.

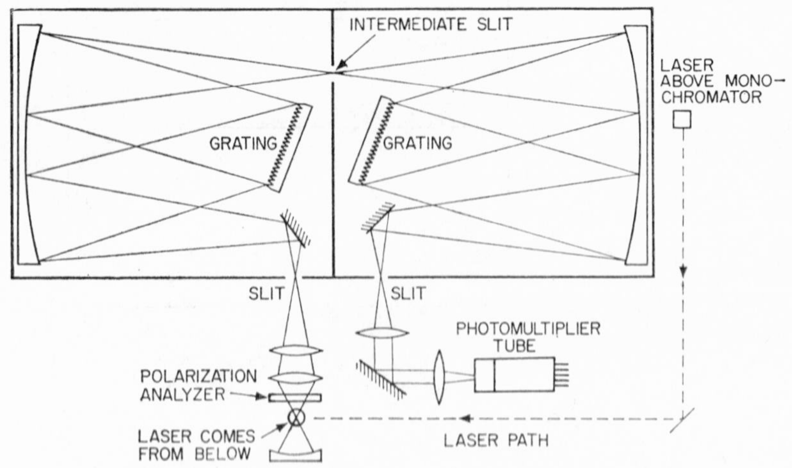

Figure 2.10 The Spectra Physics 700 spectrometer (plan)

A small (40×23 in) table-top spectrometer.

Source. Spectra Physics He–Ne type 124A; 15 mW at 6328 Å.

Sample system. Focused laser; 90° system; liquid cells available; powders and crystals accessible; analyzer and polarization scrambler incorporated.

Monochromator. Double Ebert system with gratings mounted back-to-back; 400 mm focal length; pre-set slits at 1, 2, 4, and 8 cm^{-1}; aperture *f*/5·6; gratings 1180 lines/mm, blazed at 5000 Å; range 23,500—11,300 cm^{-1}; scanning by stepping motor.

Detector–amplifier–recorder. I.T.T.-FW130 tube; cooler available at extra cost; photon counting detection plus automatic switching to d.c. at high light levels; stepping motors driven together for scanning the recorder and monochromator; recorder of *x–y* type, with 11×7 inch platten.

Price. ~25,000 U.S. dollars (~£10,000) including 124A laser; 32,850 dollars including He–Ne and 141 Ar^{+} laser.

2.4.11 *Spex Ramalog*

Spex Industries Inc.,
3880 Park Avenue,
Metuchen, New Jersey, U.S.A.

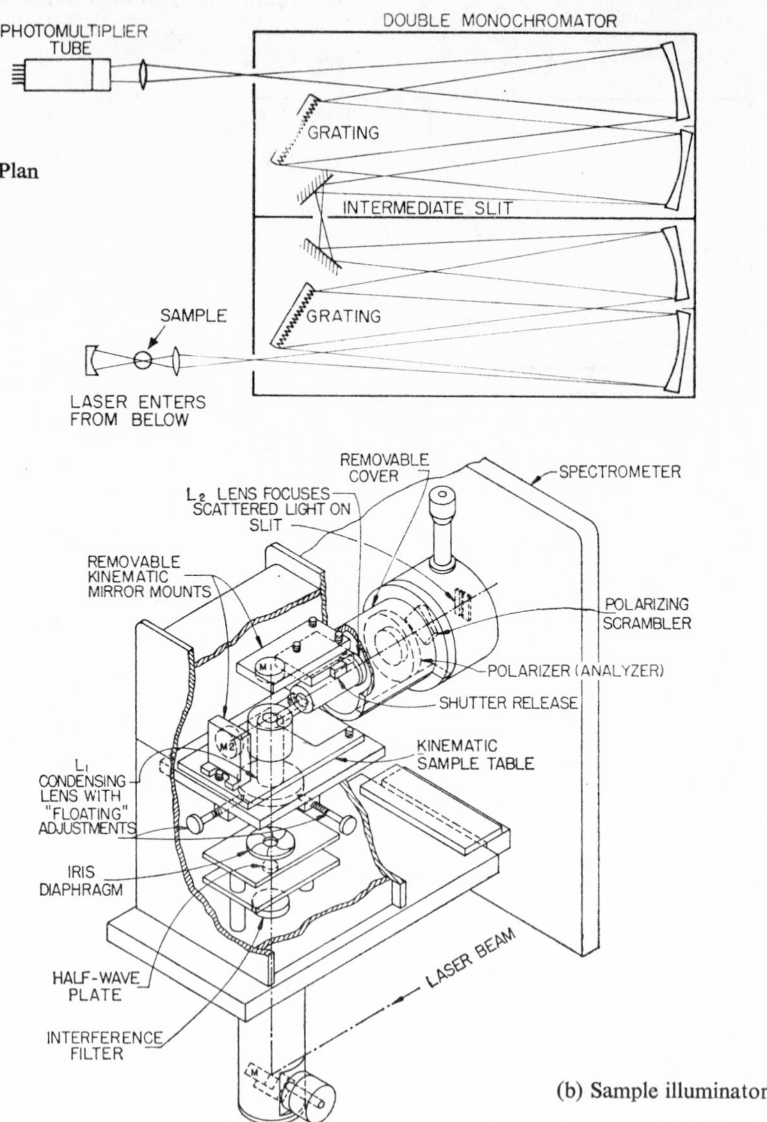

Figure 2.11 The Spex 1400 and Ramalog system (diagram by courtesy of Spex Industries Inc.)

Source. Type 125 Spectra Physics He–Ne laser is standard, but others can be incorporated at will; laser mounted beneath instrument; power-level meter swings into beam when required; interference filter fitted for powder studies.

Sample system. Beam is focused on to the sample by a microscope objective; direction of view 90°; polarization analyzer and half-wave plate incorporated, plus a wedge scrambler to guarantee accurate depolarization ratio measurements; liquid cells of volume 1 ml and 50 and 12 μl available; solids held as pellets or flat specimens; single-crystal goniometer head system available; collection aperture *f*/1·6.

Monochromator. Double Czerny–Turner instrument; aperture *f*/6·8; focal length 750 mm; resolution $\sim$0·2 cm^{-1}; gratings 1200 lines/mm; scan either manual or by motor and gearbox; scan linear in cm^{-1} or wavelength available.

Detector–amplifier–recorder. I.T.T.-FW130 type S20 photomultiplier; tube has a slit-shaped photocathode; amplification by photon-counting or d.c. amplifier system feeding a strip-chart recorder which is mechanically coupled to the wavenumber scan system.

Price. 38,000 U.S. dollars with He–Ne laser, or 52,000 dollars with Ar–Kr laser.

Experience with a Spex instrument at Southampton suggests that it is not necessary to use the manufacturer's supplied cells for liquid studies. Cells have been devised for studying samples sealed in special glass ampoules, but any convenient round or flat ampoule seems to work satisfactorily. The use of capillary tubing for powders or liquids has been fully described by the manufacturers and others in *Spex Speaker* and *Ramalogs*. It seems to be a fundamental property of the 90° arrangement that careful alignment of the sample, laser, and collection optics are essential before spectra can be recorded. The latter can be very time-consuming but rewarding! Spex Industries have recently announced a 'periscope viewer' which, when fitted to their spectrometer, facilitates alignment of the sample along the optic axis of the instrument. Recently, reports have appeared on spectra of liquids contained in thin capillaries. Good data can be recorded on samples as small as a few nanolitres.

2.5 Performance of Spectrometers

Under this heading we must consider such variables as scan speed, resolution, photometric accuracy, depolarization ratio determination, and wavelength accuracy. Unfortunately it is not possible to report on the relative merits of all the instruments available, since we have experience of only a limited number.

The better instruments are capable of providing a wavenumber repeatability and accuracy of $\sim\frac{1}{2}$—1 cm^{-1}, whilst 2 cm^{-1} should be regarded as an acceptable minimum for reliable work whether it be routine or otherwise.

Table 2.1 Raman Spectral Data for Liquid Indene (see Figure 2.12)

Identi-fication	Δν (cm^{-1})*	Line identi-fication	Neon emission line, ν (cm^{-1}) vac	Shift from He–Ne laser line, Δν (cm^{-1})
	205·0±2	1	15798·010	0
A	533·7±0·5	2	15782·381	15·6
	593·0±2		15662·306	135·7
B	730·4±0·5	3	15615·202	182·8
	831·0±1	4	15364·935	433·1
	861·0±2	5	15302·949	495·1
	947·8±1	6	15149·735	648·3
C	1018·3±0·5	7	14969·790	828·2
D	1067·8±0·5	8	14883·394	914·6
	1108·9±1	9	14427·144	1370·9
	1154·3±0·5	10	14215·950	1582·1
E	1205·6±0·5	11	13935·505	1862·5
F	1225·6±0·5		13798·503	1999·5
	1287·7±0·5	12	13439·150	2358·9
G	1361·6±0·5	13	13349·472	2446·5
H	1393·6±0·5	14	13266·384	2531·6
I	1457·6±0·5	15	12585·954	3212·1
J	1552·7±0·5		12314·085	3483·9
	1589·8±1		12287·060	3511·0
K	1610·2±0·5			
L	2892·2±1			
M	2901·2±1			
N	3054·7±1			
O	3068·2±1			
P	3112·7±0·5			

*Raman lines measured relative to neon emission spectrum.

Although this sort of accuracy is typical of reasonable quality infrared instruments in the range below 2000 cm^{-1}, it is considerably better than the performance of many in the 3000 cm^{-1} range. Owing to the short wavelength scan of a Raman spectrometer there is no justification for any fall-off repeatability over the whole range $\Delta\nu = 0$—3500 cm^{-1}.

PLATE I

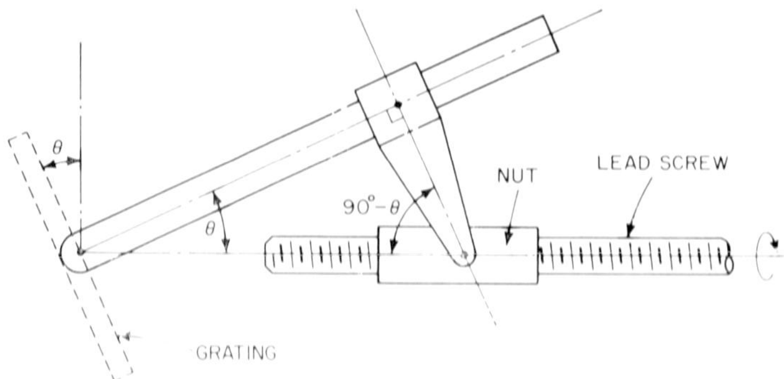

The cosecant bar drive system of the Cary 81 spectrometer; the unit coated with aluminium foil is the monochromator; the grating pivot axis is at the right of the picture; the assembly at the top left of the picture contains the controls; all of this mechanism is normally covered with panels

Also shown is a diagram of the system; the lead screw is turned at a constant angular velocity, and thus cosec θ varies linearly

[*Facing p. 50*

PLATE II

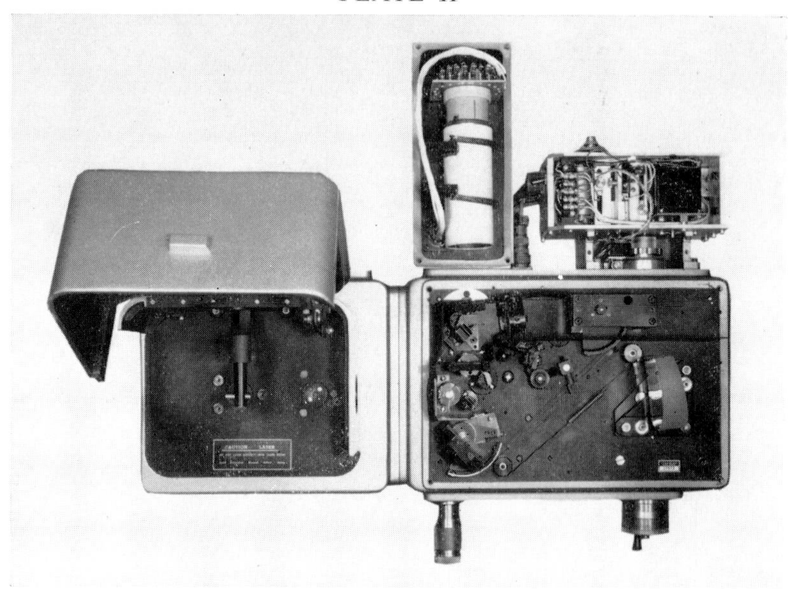

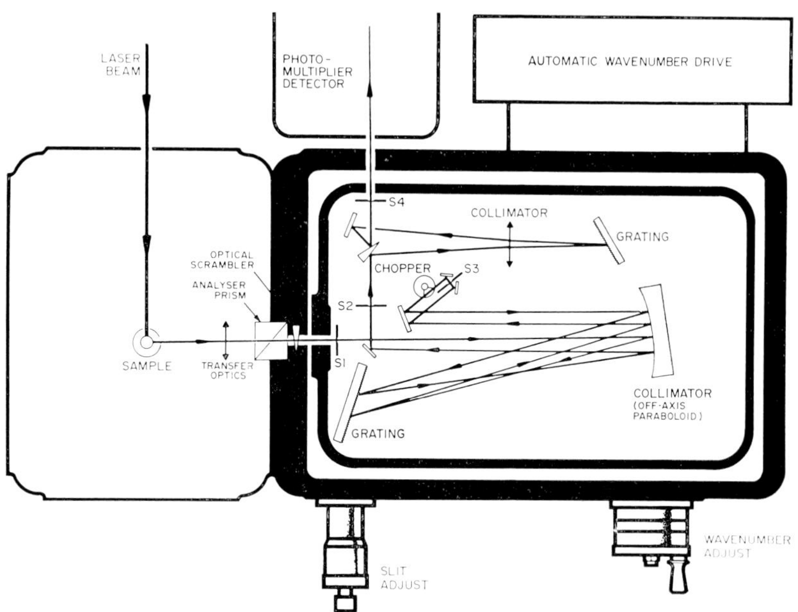

The Perkin-Elmer LR-3 spectrometer, with covers removed, viewed from above, and a diagram of the instrument

A wavelength scan range of $\Delta\nu = 0$—3500 cm^{-1} has to be regarded as the minimum acceptable. If the machine is to be designed to accept a number of laser sources this may mean that a long wavelength range must be incorporated. Some instruments will operate over a very wide range (4000—8500 Å is typical) but for an instrument designed to operate solely off the helium–neon laser a much shorter scan is required [$\Delta\nu = -1000$ (anti-Stokes) to +3500 cm^{-1} would be adequate, i.e. ~5700—8500 Å].

Calibration of the machine can be carried out in a number of ways; the use of the helium discharge lines is probably the most accurate. Hendra and Loader[45] have shown that the Raman spectrum of liquid

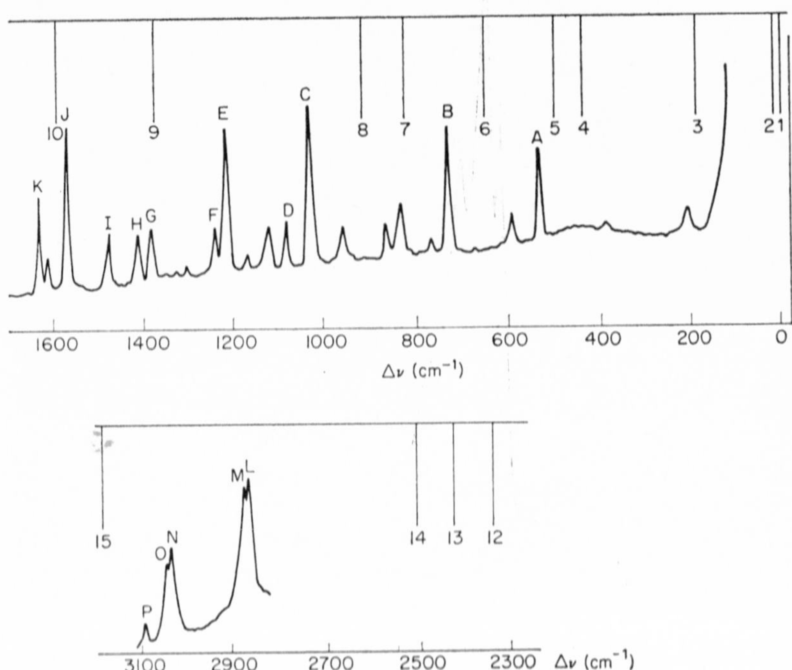

Figure 2.12 Raman spectrum of indene liquid; the letters and numbers refer to Table 2.1; the lines projected downwards are neon emissions

indene can be very useful and convenient as a calibrant. In Figure 2.12 and Table 2.1 are given the relevant data. Loader[43] has given more details of calibrations recently applied to both the He–Ne 6328 Å and Ar^{+}4880 Å source lines, and includes in his book a set of solvent spectra invaluable to the Raman spectroscopist.

As a general rule, the recording of Raman spectra with commercial equipment is about as rapid as that of infrared data. Although it is possible to record a spectrum in 5—10 minutes, best results require considerably more time. At Southampton, a scan speed in the range 1—2½ cm^{-1}/sec is considered typical unless there is some difficulty with the sample. Judging from spectra published by the various manufacturers, a signal/noise ratio of 10—50 (depending on the 'difficulty' of the sample)

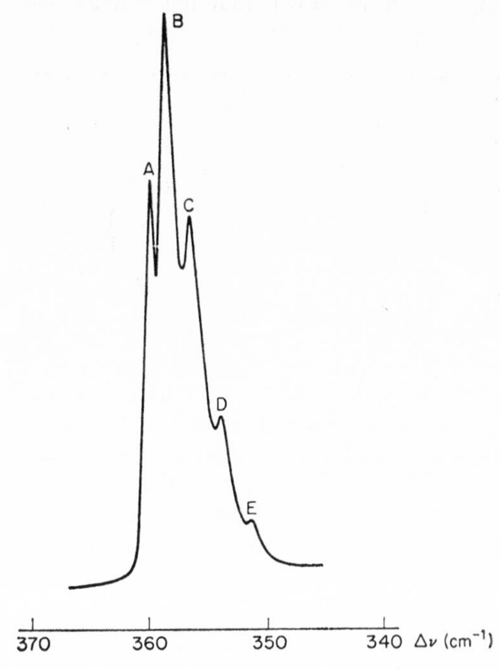

Figure 2.13 Raman spectrum of solid hexachlorodisilane (Si_2Cl_6) at 70°K; the band shown is the Si–Cl symmetrical stretch (a_g); splitting due to the ^{35}Cl and ^{37}Cl isotopes. (A) 359·9, (B) 358·3, (C) 356·5, (D) 354·0, and (E) 351·5 cm^{-1}

seems to be accepted as reasonable. Various resolution values are claimed by manufacturers (from 0·1 cm^{-1} by one to 1 cm^{-1} by another), and most of these are impossible to verify unless one is interested in gas-phase spectra. A value of 1 cm^{-1} seems quite adequate for the best condensed-phase studies even when one is interested in isotope effects, since the bands in condensed phases tend to be broad. In Figure 2.13 is shown a spectrum of hexachlorodisilane in which the chlorine isotope effect is nicely demonstrated. It can be seen that in this case a resolution of 1 cm^{-1} would be adequate. This limit is probably sufficiently stringent for single-crystal studies too.

Two other features of a Raman spectrometer which must be critically evaluated concern the discrimination of the monochromator and the recording system. Considerable experience of recording spectra at Southampton has led to the conclusion that the stray-light performance at $\Delta\nu = 100$ cm^{-1} is probably more important for most purposes than the rapidity with which the intensity of the stray light falls at small $\Delta\nu$ values. It is comparatively rare for one to be particularly interested in bands nearer than 25 cm^{-1} to the exciting line except when studying gases. In these cases the intensity ratio $I\nu/I_{\text{Raman}}$ is moderately favourable, and therefore high optical discrimination is not such an important parameter. We consider that it is essential to have an instrument with a fully coupled recorder, i.e. it must be possible to run the spectrum and return to scan interesting bands again and again and/or print onto the chart depolarization results without losing frequency calibration. This is particularly essential with Raman spectroscopy because with an emission process many different sensitivities may be required for a single spectrum. To illustrate this point a 'raw' spectrum as recorded on the Cary spectrometer is shown in Figure 2.14. It is immediately evident that the depolarization ratio of a band can be instantly assessed. Strong bands can be attenuated by known amounts so that they stay within the margins of the chart, and their strengths can be estimated; weak bands can be amplified for the same purpose. The continuous strip-chart recorder with wavenumber blip indication is as anachronistic as it is in infrared work.

Unlike infrared instruments, which have had time to evolve into a comparatively rigid 'best' design, commercial laser Raman machines have very different specifications. As a result some machines are considerably more versatile than others, while some instruments are more suitable than others for gas-phase or single-crystal studies. Fortunately the instrument manufacturers seem willing to prove their wares by recording spectra for potential customers, and so some comparison of performance is possible. However, the spectra published by the instrument makers tend to be those which are very easy to record. The following materials provide excellent tests for performance of spectrometers:

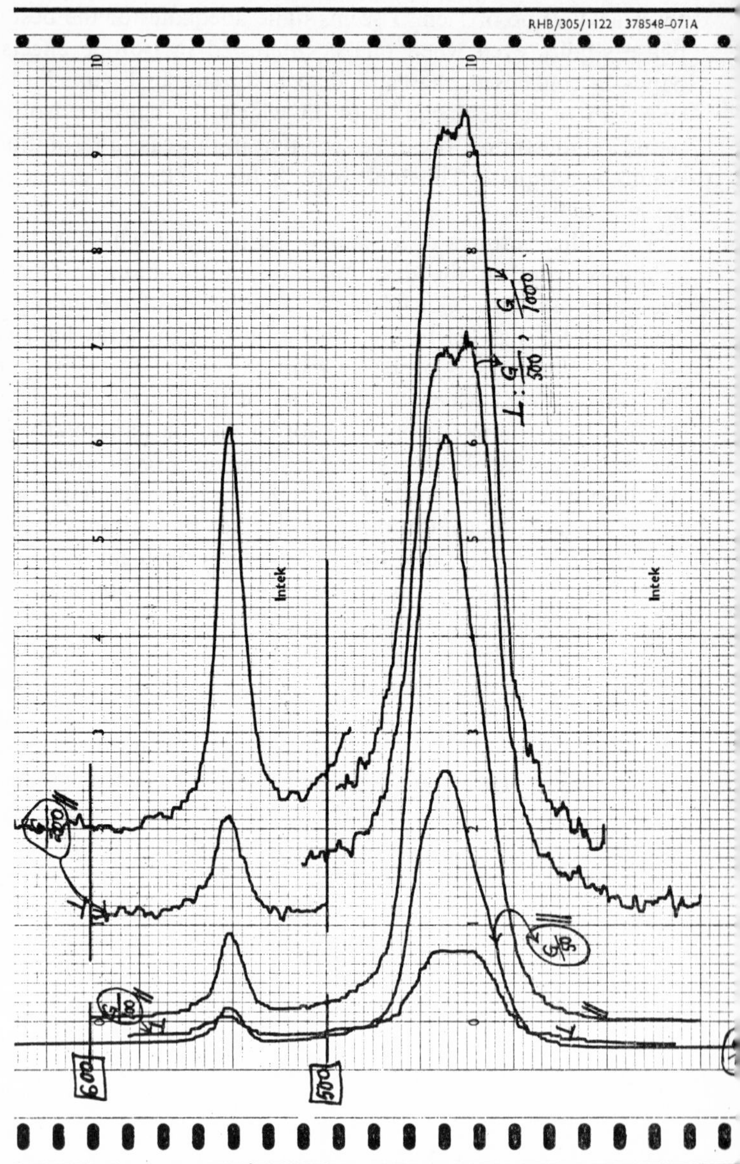

Figure 2.14 Raman spectrum (half actual size) of sulphur monochloride recorde
cm⁻¹/inch. The spectroscopist (Mr. M. Qurashi) has made notes on the chart:
and ∥ and ⊥ refer to the polarization da

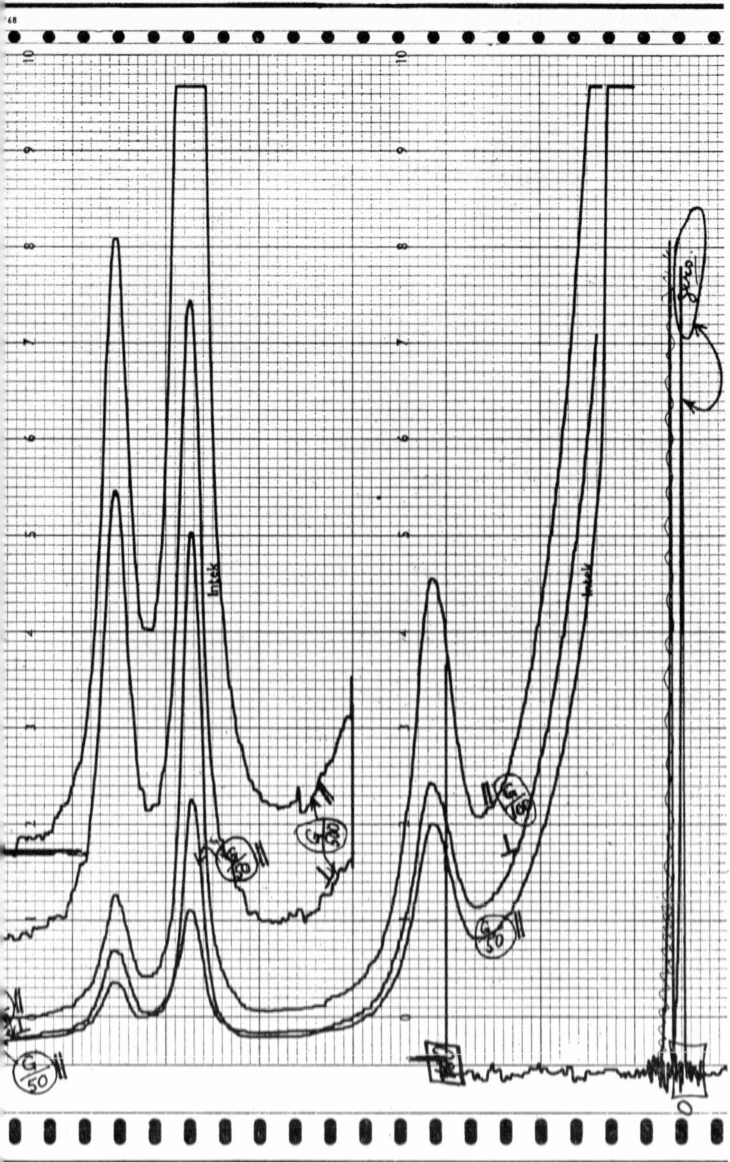

:ary 81 spectrometer; slitwidth 5 cm⁻¹; scan speed 1 cm⁻¹/sec; presentation 40 to the amplifier gain factor; the numbers in rectangles are the cm⁻¹ shifts, Δν, :t ⊥ is the *crossed* polarizer position

To indicate sensitivity	dilute solution of BF_4^-
Coloured compound	$KMnO_4$ solid and aqueous solution (He–Ne laser)
Polymer	polytetrafluoroethylene
'Difficult' powder	$BaSiF_6$
Gas	Rotation–vibration fine structure of oxygen in air

Needless to say, all manufacturers tend to claim good photometric accuracy. Although they may be strictly correct for non-polarized species, there is a significant error in some optical arrangements when scanning samples illuminated by lasers. This difficulty arises because the efficiency of a monochromator is anisotropic. The whole subject is ventilated in Chapter 3 and will not be discussed further here. However, to avoid difficulties in the determination of depolarization ratios, a polarization scrambler should be incorporated before the entrance slit of the monochromator if an analyzer is either not present or has to be turned.

The only absolute intensity measurements to date using laser sources are those by Damen, Leite, and Porto,[39] but there is no reason to believe that there will not be a proliferation in the near future. In principle it is much easier to estimate the radiation intensity for a laser than for a discharge lamp because the beam is well collimated. It is also possible to make intensity measurements with 180° excitation using an external-standard method developed recently at Southampton. The cell, developed by Qurashi, Hendra, and Mackenzie,[46] is shown in Figure 2.15. The laser

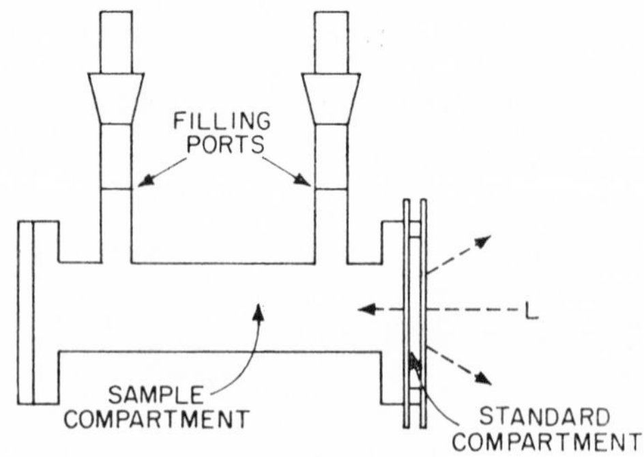

Figure 2.15 A quantitative cell for the Cary 81 spectrometer

passes through both the sample and the standard liquid cavity. The length of this cavity can be varied so that the intensity of the standard spectrum can be kept similar to that of the unknown. Any liquid whose Raman spectrum has been examined with a view to determining the absolute band intensities can be used as a standard. Again the whole problem of intensity measurement is considered in more detail in Chapter 3.

No commercially available instrument satisfies all the criteria for the ideal Raman spectrometer. Some come close to this situation but they tend to be rather expensive. In general, it seems that for chemical applications the machines available have to be compared in terms of their sampling arrangements and versatility. As far as published spectra are concerned, there seems to be insufficient evidence to enable one to classify the available instruments with regard to their monochromator performance or detector–amplifier–recorder systems (except in one important respect, i.e. only one manufacturer offers a machine fitted with a completely coupled recorder). However, it seems that the claims made for detector cooling are sufficiently vociferous to interest even the sceptic, and therefore the inclusion of this feature must at least be regarded as desirable. It is worth noting that statistics require that the noise at the detector should be proportional to $(\text{signal})^{\frac{1}{2}}$. As a result there is a limiting light level above which the fundamental noise from statistical variation in the source exceeds the dark current of the tube. It seems probable that, for condensed-phase spectroscopy, the background always gives rise to sufficient noise to make detector cooling redundant. For gas-phase spectroscopy this is not so, and detector cooling may be of value.

Turning to sample geometries, the two principal contenders are the 180° Cary type system and the 90° arrangement common to all available machines. Experience has shown that the 180° system has a number of advantages as follows:

(a) It is very convenient to examine samples in whatever shape or form they may be. No sample preparation (such as formation of pellets from powders) is required.

(b) No optical alignment is needed, in most cases, before the spectrum is recorded.

(c) It is possible to study powders, liquids, and particularly crystals, inside their glass containers simply by 'looking' through the wall of the vessel. This is essential where the compounds are sensitive to air or moisture, or where powders are too unstable to be made into pellets. The 90° system may not be quite so convenient or versatile in this respect.

(d) Owing to the small degree of penetration required for this geometry, opaque samples and surface films tend to be readily accessible as samples, but this advantage is shared to some extent by the 90° system when examining species held at 45° to the illuminating and viewing directions.

The 90° arrangement has the following points in its favour:

(a) It is possible to record spectra closer to the exciting line. In the 180° system, difficulty in this respect is caused, at least partially, by the collection of Raman light emitted by the collector lens and cell windows where they are traversed by the laser.

(b) By use of suitable cells (usually of fairly large volume), depolarization ratios of very high accuracy can be obtained.

(c) 90° excitation and the very large sample areas offered by some manufacturers make it very easy to accommodate bulky experiments of a specialized nature.

(d) Gas-phase samples are accessible to the better instruments incorporating this form of excitation.

(e) The minimum sample size for 90° excitation is extremely low. Liquid-phase spectra have been recorded on 8 nanolitres of liquid in a capillary.[47]

The two techniques have recently been compared by Haber and Sloane[48] of the Cary company. In addition, it should be noted that some crystal data can be collected only if both the 90° and the 180° system are available (see Chapter 4).

When determining Raman spectra of absorbing (coloured) materials it is customary to use the Lippincott–Fisher relationship.[49] This, in effect, balances the absorption of the radiation caused by increasing the concentration of a solution against the increased intensity of the Raman lines resulting from greater concentration. Gall[50] has studied the problem as it applies when 180° excitation is used. His results confirm that there is an optimum concentration of an absorbing solution from a Raman point of view.

High-speed Recording of Raman Spectra

By use of laser sources it is possible to record Raman spectra at very high speeds. Since the detectors used in the visible are of the 'quantum' rather than the 'thermal' type, it is intrinsically much simpler to record data rapidly in the Raman effect than in the infrared. As speed rises, the signal strength has to be increased to maintain an acceptable signal/noise ratio, and thus the potential and proved advantages of laser sources for this type of work are apparent. A recent review by Delhaye[51] explains the instrumental problems and their solution, and gives a number of examples of spectra recorded at very high speeds.

Delhaye and his co-workers have used a number of approaches as follows:

(1) *A modified Coderg PH-1 instrument.* A cam-operated scan of wavelength is used with a cooled detector reading into an oscilloscope. A speed of 1000 cm^{-1}/sec has been achieved with this arrangement, with a resolution of 5 cm^{-1}. Spectra of chlorine in arsenic trichloride solution (showing the isotope 'split' triplet due to ν_{Cl-Cl}) and of azobenzene have been obtained. With an argon ion laser of 150 mW output rather than the more normal helium–neon device, the products of the reaction $HgCl_2+HgBr_2 \rightleftharpoons 2HgClBr$ have been studied at 500 cm^{-1}/sec.

(2) *Image intensifiers in a spectrograph.* The advantage with this approach is that scanning is avoided (with consequent reduction in the mechanical complexity of the instrument), and therefore the whole of the spectrum is accumulated by the detector during the whole 'exposure'. The use of image intensifiers is due to Bridoux[52]. A grating spectrograph is fitted with commercial image intensifiers (of S20 type) in place of the photographic plates, and the spectrum is displayed with a photon gain of 10^4—10^5 on a phosphor screen in the green. The gain is $\sim$1000 times that for the best photographic plates sensitive in the red. The spectrum is recorded with a conventional camera.

(3) *The Vidicon television camera tube.* This consists of a selenium photoconductive layer upon which the picture is produced, and an electron gun system which senses the degree of conductivity of the layer. The spectrum can be focused on to the sensor of the tube, or the tube can be used as a secondary detector after the image intensifier described above. The advantage of the Vidicon tube is its integrating property (in which it resembles a photographic emulsion). The electrical signal is displayed on an oscilloscope.

Delhaye has given numerous examples of his work, but only two will be mentioned here:

(a) A dilute solution of azobenzene (M/100); spectrum of good quality recorded in 1 millisec using a He–Ne laser of 160 mW power.

(b) A spectrum of *o*-dichlorobenzene of superb quality recorded using a single pulse from a Siemens ruby laser (0·01 J) in 1 millisec. The frequency range is $\sim$250 cm^{-1} wide.

Finally, a particularly elegant experiment must be described since it demonstrates the potential of this approach. The reactions $2NOBr \rightleftharpoons 2NO+Br_2$ and $2NOCl \rightleftharpoons 2NO+Cl_2$, in which the reactants have half-lives of from 1 to 1000 sec in the gas phase, were examined. A single flash

0·5 millisec in duration from a ruby laser was used, photodecomposition was initiated, and reaction demonstrated by the appearance of a doublet at 548 cm^{-1} due to chlorine. A signal/noise ratio for this feature of ~20 was achieved.

3

Raman intensities and depolarization ratios

In Chapter 1 we surveyed briefly the quantum and classical approaches to Raman intensity. We now elaborate on those ideas and relate them to some of the work in the literature. This is a rapidly expanding field, and no definitive account could be written at the moment. Our intention is to provide a guide to the literature.

We shall first relate the material of Chapter 1 to the Raman spectroscopy of fluids, or more strictly to idealized random arrays of molecules. Having presented the mathematical formulae relevant to this situation, we shall go on to consider measurements on real systems with real instruments.

The reader will probably be familiar with the occurrence of 'polarized' and 'depolarized' Raman bands in fluids. Information is also obtainable from the relative intensities of different bands in a spectrum, and the actual value of the 'depolarization ratio' for polarized bands (Chapter 3.1.1).

Because of their directional properties, lasers make it easier, in principle, to obtain accurate intensity and, particularly, depolarization data than with discharge lamps. This is true both for specialists who build their own apparatus and work to very fine limits, and for users of commercial instruments. Unfortunately detailed treatments of the correction factors applicable to laser systems have not yet appeared, and even the theory of ideal measurements has been misunderstood.[53] We therefore consider a number of ideal systems, and then present a few tentative ideas on the probable departure from ideality of real versions of these. We use, in part, the generalizations derived by M. J. Gall[50] from the previous treatments of Placzek,[3] Woodward and Long,[54] and Wilson, Decius, and Cross.[2]

3.1 Treatment for Fluids

A rigorous treatment can be derived for a perfectly random collection of non-interacting molecules. No liquid satisfies the latter criterion, and some anomalies are therefore apparent. It is important to distinguish between permanent (ground state) interaction with neighbouring molecules, and interference with the scattering process by neighbouring molecules. A correction must be made for the latter before anything can be said about the former. This point is taken up again later, after we have developed the theory of measurements.

The treatment of scattering in Chapter 1 concerned single molecules only. 'Molecular crystals' are treated for most purposes as an array of oriented molecules, and insofar as this approximation is valid, the molecular Raman tensor components can be related to the crystal Raman tensor components in an obvious way. This point is developed in Chapter 4 which deals with single-crystal spectroscopy.

In fluids much of the information is lost because of the random orientation of the molecules. We now consider what may be salvaged. Accurate observations on fluids are greatly facilitated owing to their completely isotropic optical behaviour and freedom from macroscopic flaws. We attempt here to use general principles, which may if necessary be applied by the reader to novel geometries. We also show the application to some of the more common geometries in present use.

We have noted that the description of a tensor property varies according to the coordinate system used. However, there are two *invariant* quantities which are independent of the choice of axes: the 'mean value' and the 'anisotropy'. We illustrate with the polarizability tensor. For the mean value:

$$\bar{\alpha} = \tfrac{1}{3}(\alpha_{xx}+\alpha_{yy}+\alpha_{zz}) \tag{1}$$

whilst the anisotropy $\gamma(\alpha)$ is given by:

$$\gamma^2(\alpha) = \tfrac{1}{2}[(\alpha_{xx}-\alpha_{yy})^2+(\alpha_{yy}-\alpha_{zz})^2+(\alpha_{zz}-\alpha_{xx})^2+6(\alpha_{xy}{}^2+\alpha_{xz}{}^2+\alpha_{yz}{}^2)] \tag{2}$$

The common notation for the anisotropy is γ, but we introduce the new notation γ(tensor) to distinguish from the second hyperpolarizability and between the anisotropy of $\boldsymbol{\alpha}$, $\partial\boldsymbol{\alpha}/\partial q_i$, and $\boldsymbol{R}$.

It is clear that, for a fluid, we cannot predict the orientation of a scattering molecule. However, the (symmetric) tensor nature of the scattering parameter leads to certain directional properties of the scattered light even when random orientation of the molecules is considered. These directional properties are expressed in terms of the *average* scattering tensor, for which we find:

$$\overline{\alpha_{ii}^2} = \overline{\alpha_{xx}^2} = \overline{\alpha_{yy}^2} = \overline{\alpha_{zz}^2} = (1/45)[45\bar{\alpha}^2 + 4\gamma^2(\alpha)] \qquad (3)$$

$$\overline{\alpha_{ij}^2} = \overline{\alpha_{xy}^2} = \overline{\alpha_{xz}^2} = \overline{\alpha_{zy}^2} = (1/15)\gamma^2(\alpha) \qquad (4)$$

and similarly for the Raman tensor.

We are, of course, at liberty to fix the (orthogonal) axes of the average tensor anywhere, so we choose directions ('laboratory axes') which are important in a particular scattering geometry. The implication of the average tensor is that the observed scattered light has the directional characteristics of light from an *oriented* scatterer with that scattering tensor. We may consider in particular 'isotropic scattering' governed by the first term $\overline{\alpha_{ii}^2}$, in which scattered light is generated by an oscillating dipole in the same direction as the electric vector of the exciting light, and 'anisotropic scattering' governed by the second term $\overline{\alpha_{ij}^2}$, in which the electric vectors generating the scattered light are in the plane perpendicular to the vector of the exciting light. The validity of the directional predictions arising from this averaged tensor has been carefully tested by Porto and his co-workers using a laser source.[55] In particular, Porto finds that, for normal excitation intensities, the scattered intensity and depolarization are quite independent of the propagation direction of the exciting light, and depend only on the direction of its electric vector with respect to the analyzer and monochromator. This is, of course, after correcting for variations in the illuminated volume viewed by the detector. This is the result expected for incoherent scattering. However, see ref. 56.

We consider now a number of simple idealized scattering geometries. Unfortunately real arrangements always depart from these ideals, because the ideals deal with a single direction of propagation for the exciting light and a single direction of propagation for the scattered light. Whilst the first ideal may be realized with a laser, the second never is, because an infinitesimal solid angle cone of radiation will always contain an infinitesimal proportion of incoherent scatter. Little attention has been given to the design of a system which would collect the maximum light with minimum loss of information. It is usual to calibrate a given set-up to allow for these errors. Such calibration is not without its own problems, which we summarize later. We also discuss the analytical form expected for the errors.

In discussions of discharge sources, it is commonly stated that 'convergence errors' arise, being sometimes defined as all errors which appear as a result of departure from the exact right-angle between propagation directions of incident and scattered light. However well this formulation may be justified with regard to discharge lamp excitation, it is misleading when applied to lasers. As we have seen, it does not matter in theory what relationship exists between exciting and scattered light propagation directions. Errors arise only when the relationship of the electric vectors is ill-defined, with a range of values. With discharge sources, this results from the nature of both the source emission and the collection optics. With laser sources the error arises only from the *divergence* of the Raman light. In this book we therefore refer to 'divergence errors'. As will be seen, these are a little more tractable than the double errors arising with discharge lamps.

In the following treatments we use E_x, E_y, and E_z to denote components of the (r.m.s.) exciting vector, and I_x, I_y, and I_z for the observed intensities, with x, y, or z respectively as the electric vector direction.

90° *Scattering*

(i) *Polarized exciting light* This is the conventional laser-excited set-up.

$$E_x \neq 0$$

$$E_y = E_z = 0$$

Two cases arise:

(a) *observation along x*

$$I_y = k\,\overline{\alpha_{ij}^{\,2}}\,E_x^{\,2} \tag{5}$$

$$I_z = k\,\overline{\alpha_{ij}^{\,2}}\,E_x^{\,2} \tag{6}$$

(b) *observation along y*

$$I_x = k\,\overline{\alpha_{ii}^{\,2}}\,E_x^{\,2} \tag{7}$$

$$I_z = k\,\overline{\alpha_{ij}^{\,2}}\,E_x^{\,2} \tag{8}$$

(ii) *Natural (depolarized) exciting light*

$$E_x = E_y \neq 0\ (= E,\ \text{say})$$

$$E_z = 0$$

Observations along x and y are now equivalent. Considering:

(a) *observation along x*

$$I_y = k\,\overline{\alpha_{ii}^{\,2}}\,E_y^{\,2} + k\,\overline{\alpha_{ij}^{\,2}}\,E_x^{\,2} \tag{9}$$

$$I_z = k\,\overline{\alpha_{ij}^{\,2}}\,E_x^{\,2} + k\,\overline{\alpha_{ij}^{\,2}}\,E_y^{\,2} \tag{10}$$

On-axis 180° (Cary 81) or 0° (Jarrell-Ash 25-300) scattering

(iii) *Polarized exciting light*

$$E_x \neq 0$$

$$E_y = E_z = 0$$

Collection along z

$$I_x = k\overline{\alpha_{ii}^2}\,E_x^2 \tag{11}$$

$$I_y = k\overline{\alpha_{ij}^2}\,E_x^2 \tag{12}$$

(iv) *Depolarized exciting light*

$$E_x = E_y \neq 0$$

$$E_z = 0$$

Collection along z

$$I_x = k\overline{\alpha_{ii}^2}\,E_x^2 + k\overline{\alpha_{ij}^2}\,E_y^2 \tag{13}$$

$$I_y = k\overline{\alpha_{ij}^2}\,E_x^2 + k\overline{\alpha_{ii}^2}\,E_y^2 \tag{14}$$

We discuss first the 90° scattering of unpolarized light, because the conventions relate to it. It will be apparent that two measurements may in general be of interest. One is the state of polarization of the scattered light; the other is the relative intensity of different Raman bands. We shall later relate these to the invariants of the scattering tensor for various scattering geometries. It should be obvious at this stage that *observed* relative intensities may depend not only on scattering geometry but also on the polarization of the measured bands.

3.1.1 *Polarization*

The polarization of Rayleigh and Raman light is usually expressed as a 'degree of depolarization' or a 'depolarization ratio', given the symbol ρ. For case (ii)(a), clearly $I_y > I_z$, so we express this ratio as I_z/I_y which will have the value zero for completely polarized scatter. Then for Rayleigh light:

$$\rho_n = \frac{I_z}{I_y} = \frac{6\gamma^2(\alpha)}{45\bar{\alpha}^2 + 7\gamma^2(\alpha)} \tag{15}$$

where 'n' stands for natural excitation.

It may be seen for a completely isotropic molecule that $\gamma(\alpha) = 0$, so that $\rho_n = 0$. It is also apparent that $\rho_n < 6/7$ in all cases since $\bar{\alpha}$ may not be zero. (Detailed treatment[6] shows $\rho_n \leqslant \frac{1}{2}$.)

The case of Raman scattering may be examined by putting $\bar{R}$ and $\gamma(R)$ in place of $\bar{\alpha}$ and $\gamma(\alpha)$. The only difference this makes is that $\bar{R}$ may be zero, since the components of $\boldsymbol{R}$ (or of $\partial\boldsymbol{\alpha}/\partial q_i$) may be zero or negative. Thus:

$$6/7 \geqslant \rho_n \geqslant 0 \tag{16}$$

A most important consideration is that, for non-totally symmetric vibrations, $\bar{R} = 0$, so that $\rho_n = 6/7$. Correspondingly, for totally symmetric vibrations $\bar{R} \neq 0$, so that $\rho_n < 6/7$. For example, in molecules belonging to cubic point-groups *only*, for totally symmetric vibrations $\gamma(R) = 0$ so $\rho_n = 0$, as with Rayleigh scattering from such molecules. These considerations are probably familiar to the reader; in principle, Raman bands may be assigned to totally symmetric or other vibrations by measurement of the depolarization ratio. The only precaution to be observed is that this ratio may occasionally approach very closely to 6/7 for some totally symmetric bands. An internal standard may sometimes be used for the detection of such bands.[57]

For pure rotation it is found that only the anisotropy may be non-zero and so contribute to Raman activity. Thus, for pure rotational bands $\rho_n = 6/7$ in all cases. This is understandable classically, since the activity is due entirely to the change of aspect of the molecule.

Returning to the case of polarized exciting light, it can be seen that the basic requirement is to make measurements which depend differently on the isotropic and anisotropic parts of the averaged tensor. For 90° scattering there are two possible schemes which will accomplish this. One is to distinguish I_x and I_z in case (i)(b) for which:

$$\rho_l = \frac{I_z}{I_x} = \frac{3\gamma^2(\alpha)}{45\bar{\alpha}^2 + 4\gamma^2(\alpha)} \tag{17}$$

where 'l' stands for linear. Clearly, for Raman scattering:

$$0 \leqslant \rho_l \leqslant \tfrac{3}{4} \tag{18}$$

with similar considerations for the symmetry of the vibration to those applying to ρ_n.

The second scheme is to take one observation along x and another along y. Experimentally this would be accomplished by changing the polarization of the exciting light rather than by swinging round the monochromator. This is in fact the normal procedure for both laser and discharge-lamp excitation. For helical and similar arcs (e.g. the Toronto Arc) this scheme has the great advantage that illumination of the sample from *all* directions at right angles to the monochromator axis is permissible, and indeed advantageous, the light being polarized in the Edsall method[58] either parallel or perpendicular to this axis with a Polaroid sheet cylinder. For such illumination, using no analyzer:

$$\rho = \frac{I_{\text{axial}}}{I_{\text{cross}}} = \frac{I_y + I_z \text{ of (i)(a)}}{I_x + I_z \text{ of (i)(b)}} = \rho_n \tag{19}$$

This is therefore an alternative and practical method of observing ρ_n. For a laser source, the polarization of a polarized laser beam can be turned with a half-wave plate. When using this scheme with the laser, it must be remembered that a normal monochromator transmits light preferentially for one polarization. This difficulty is avoided with cylindrical illumination in some arcs. For use with the laser there are two possibilities. The Raman light must either be depolarized with a polarization scrambler, leading to observation of ρ_n, or definitely analyzed for that part with the electric vector perpendicular to the laser propagation direction, leading to observation of ρ_l. This difficulty with the monochromator also, of course, applies to direct observation of the state of polarization of the Raman light. If it is desired to make the measurement directly, either the polarization must be scrambled after passing through the analyzer, or a correction factor must be applied. The latter method is useful only where the factor is close to unity; it will anyway vary with the absolute frequency of the Raman line.

It is apparent that, for on-axis scattering, an analyzer and polarized excitation must be used to observe the depolarization ratio. Then ρ_l is observed. The difficulty with the monochromator may be overcome with a scrambler, or, better, by turning only the exciting vector. This is customary, and it is accomplished with a polarized laser and a half-wave plate. The half-wave plate is always left in the optical train to prevent disturbance by reflection losses.

The discussion so far has effectively considered idealized versions of all currently available scattering geometries. The reader may care to speculate on variations, or apply similar reasoning to other suggested geometries. For instance, the exact reverse of the arc system, with 90° collection round the entire circle in the plane perpendicular to the laser, would be theoretically advantageous if it could be realized.

3.1.2 *Intensities*

The measurement of absolute intensities is a very complicated process, and it is not reasonable to describe the details in an introductory text.

However, the reader may wish to use some technique which requires the measurement of relative intensities. This is clearly the place to point out some of the geometrical hazards involved. For completeness we consider quite briefly the other major instrumental and theoretical corrections involved. For fluids and powders the relative intensities of two different bands will be involved. In Chapter 4 we consider the precautions necessary in measuring the relative intensity of the same band in different orientations of a crystal. This is comparable to the observation of depolarization ratios in fluids.

We shall consider the scattering arrangements listed previously, and relate them to the averaged tensor and hence to the invariants. From the previous discussion it will be apparent that, in order to obtain meaningful results from scattering with a laser source, the combined monochromator and detector polarization discrimination must either be known at the absolute frequencies involved, or overcome with a scrambler or an analyzer. This is true for polarized or unpolarized lasers. This precaution may be neglected only for excitation with cylindrical symmetry about the monochromator axis, as with a helical arc or an unpolarized on-axis laser. With a polarized laser it clearly only makes sense to collect in a direction perpendicular to the exciting vector, if polarized bands are not to be lost or much attenuated. We consider the previous case (i)(b):

$$E_x \neq 0$$

$$E_y = E_z = 0$$

90° collection along y

$$I_x = kR_{ii}^2 E_x^2 = (kE_x^2/45)[45\bar{R}^2 + 4\gamma^2(R)] \tag{20}$$

$$I_z = kR_{ij}^2 E_x^2 = (kE_x^2/15)\gamma^2(R) \tag{21}$$

Clearly, if an analyzer is to be used it must analyze for I_x, or again some bands may be lost or attenuated. On the other hand, if a scrambler is used:

$$I_x + I_z = (kE_x/45)[45\bar{R}^2 + 7\gamma^2(R)] \tag{22}$$

With an unpolarized laser we see that if an analyzer is used it must similarly be set to pass the component with vector perpendicular to the propagation of the exciting light. The possible simple arrangements are summarized in Table 3.1. Some caution may be necessary in using this table. Optical elements must not be added to the train between measurements since losses are always involved. A change from 90° to on-axis collection (possible in the Cary 81) is accompanied by a change in collection efficiency (i.e. k is changed). The 'scrambling' of a laser (vector E) leads, even with a 'perfect' scrambler, to $E_x = E_y = E/\sqrt{2}$. Finally, interchange even of a scrambler and an analyzer of identical transmission characteristics (for the appropriate polarized light) is not permissible, since k is again effectively changed in the monochromator.

The entry in Table 3.1 appropriate to the helical and annular discharge lamp arrangements is also noted.

Table 3.1 Summary of Some Possible Relative Intensity Measurements using Clear Fluids and Ideal Scattering Systems

Exciting vector direction (r.m.s. value E)	Analyzer, and collection	Observed intensity* $\times 45/kE^2$
x	x, 90° or 180°	$45\bar{R}^2 + 4\gamma^2(R)$
x	scrambled, 90° or 180°	$45\bar{R}^2 + 7\gamma^2(R)$
$x + y$	x, 90° or 180°	$45\bar{R}^2 + 7\gamma^2(R)$
$x + y$	scrambled, 90°	$45\bar{R}^2 + 13\gamma^2(R)$†
$x + y$	no analyzer, 180°	$45\bar{R}^2 + 7\gamma^2(R)$
x	no analyzer, 90°	$45\bar{R}^2 + (4 + 3m_\nu)\gamma^2(R)$‡
$x + y$	no analyzer, 90°	$45\bar{R}^2 + (7 + 6m_\nu)\gamma^2(R)$‡

*In general k is different for each entry.

†This entry is equivalent to the helical or annular arc arrangement with neither polarizer nor analyzer.

‡m_ν = (Instrument response for second polarization)/(Instrument response for vector along x) at frequency ν.

Part of the object of Table 3.1 is to show that, if relative intensities from two different experiments are to be compared, allowance must be made for the anisotropies of the tensors (depolarizations) for the bands concerned. This would apply, for instance, where arc and laser measurements are to be compared.[53] In any interpretation of relative intensities it is clearly important to divide the observed intensity into parts due to $\bar{R}$ and to $\gamma(R)$. This is done by observing the depolarization ratio, and applying the relevant formulae.

Other Factors. We have already considered the different instrument response to different polarizations. This may be determined with an ordinary tungsten lamp operated off a stabilized supply. Such a source is unpolarized, and Polaroid sheet may be used to provide horizontally and vertically polarized light of equal intensity. The polarization factor and the absolute response vary with frequency; to make accurate relative measurements at widely separated frequencies, or absolute measurements, the response must be determined with a tungsten band lamp or a reference sample.[59,60] The monochromator dispersion also varies with frequency, so the same geometrical slit represents a different spectral slit. The slit must also be sufficiently narrow to give an accurate measurement.

We cannot go further into the details of these corrections, but we hope that the interested reader will now be in a position to know what information must be sought from specialist sources.

3.1.3 *Geometrical–Optical and Divergence Effects*

We now consider some of the ways in which practical systems depart from ideality, drawing a clear distinction between effects on the relative proportion of isotropic and anisotropic scatter, which must always be considered, except when studying the totally symmetric vibrations of molecules with cubic point-groups, and other effects which need only be considered when different samples are compared. It is unfortunate that many of the most efficient collection systems are those which are most subject to loss of ideality.

By geometrical–optical effects we mean variations in collection efficiency and observed isotropy/anisotropy contributions arising from change of sample refractive index. We have mentioned also that all discharge-lamp systems are subject to a 'convergence error' because the propagation direction of the emission is not well defined. In addition, for the most efficient systems the propagation direction of the Raman light may be very poorly known. For instance, in the Cary 81 Toronto Arc 7 mm diameter and capillary cell systems, much of the light is collected by internal reflections off the cell walls. Even in the 19 mm diameter cell, where such reflections do not contribute to observed intensity, the finite aperture results in some loss of precision. This we term 'divergence error', since it arises from the divergence of the Raman light.

The corrections applicable to discharge lamps are discussed by Brandmüller and Moser[61] and others,[54,60,62,63] and are summarized by Hester.[4] Clearly, only the divergence error for the Raman light remains in a well designed laser system, although in very careful work it is necessary to avoid stress depolarization of the laser beam in optical components.[64]

Specialist workers requiring very accurate results invariably construct systems in which the errors are either clearly defined or reduced to the point where they may be neglected for the accuracy required. Some of the papers already cited[55,64] are good examples of this approach. The use of lasers has made such work possible for the first time. Thus Murphy *et al.*[64] were able to study the very small depolarization ratio ($\sim 5\times10^{-3}$) of ν_1 of carbon tetrachloride in various environments. This departure from the ideal value of zero is thought to be due principally to intermolecular interactions in the liquid phase, rather than to isotopically substituted species. The depolarization ratio for Rayleigh scattering in a number of pure liquids has also been studied.[38]

We shall not be further concerned with very accurate determinations. We now consider the less than ideal systems which are available on commercial instruments and which are perfectly adequate for most purposes. The convergence of the exciting light is no longer a problem with laser sources, even in the least ideal systems. Although the laser is often focused into the sample, the angle of convergence is still fairly small, e.g. up to about 3° between extreme rays, or $f/15$ for a 5 cm focal length lens (as on the Spex), or about 1° ($f/70$) for a 20 cm lens in the Cary 81.

The most efficient collection systems utilize multiple reflection of the Raman light off the walls of the cell. The effect of this is not really open to detailed analysis, and such systems are never used where samples of varying refractive index are to be compared. Rea[62] has reported some empirical studies which show the expected sharp drop in collection efficiency where the sample and the cell have the same refractive index. Multiple reflection of the Raman light is also used to advantage in the Cary laser on-axis capillary cell system. New errors arise in that case because the depolarization of the Raman light is measured as such, but is partially lost during the reflections. However, it is still possible to tell the difference between polarized and depolarized bands quite easily in most cases. This failing is a characteristic of all systems in which the sample is retained in a capillary, but they are useful for very small samples.

The Cary on-axis system can be used for larger samples without a capillary cell, but this results in much lower efficiency, particularly for clear specimens (up to 50-fold reduction in observed intensity). The most efficient arrangement for clear samples, apart from the on-axis system with capillary cell (which appears to be roughly comparable), uses multiple reflection of the laser beam. This can be achieved with a special wedge-shaped cell such as that supplied with, for example, Perkin-Elmer instruments and, latterly, as a modification for the Cary 81 (see Figure 2.2). With this arrangement the laser beam is not focused, but some loss of polarization can occur on reflection. An alternative is the 'external resonator' arrangement, which can be used easily with corrosive and air-unstable samples. In this system the beam is multiply reflected between an external dielectric mirror and the laser front mirror.

Clearly, no system involving multiple reflection of the laser or the Raman light can be used in a direct (uncalibrated) comparison of intensities in different samples. Systems not involving such reflection are much more tractable, and due allowance can be made for the optical factors.

The theory for discharge-lamp excitation has been worked out on the basis of the 'Nielsen conditions',[65] which require that Raman light is collected directly from a volume entirely contained within a sample uniformly illuminated perpendicular to the monochromator axis. According to this theory, the length of the 'Nielsen prism', and therefore the collection efficiency, is proportional to the refractive index of the sample. However, Wall and Hornig[63] have pointed out that the Nielsen prism is already as long as the 19 mm diameter (non-internally reflecting) sample tube on the Cary 81 arc system, for a refractive index of unity. Thus, no such correction applies in that case. Laser systems, which require quite different analysis, have not yet been subjected to such detailed treatment. We first distinguish geometrical–optical effects as usually defined, i.e. the dependence of collection efficiency and isotropy/anisotropy contributions on sample refractive index; we later discuss departures from ideality caused by excessive divergence of the Raman light.

We wish to make only tentative suggestions regarding geometrical–optical effects. In view of the nature of laser excitation, it seems realistic to consider any laser beam cross-section as a point scattering source particularly when the beam is focused. Since Raman scatter is incoherent, we might expect that observed intensity could be obtained by integration over an appropriate beam-length. In this way we might at the very least obtain an expression in which instrument parameters could be treated as 'fudge factors' to obtain a good fit for general use. Referring to Figure 3.1, we consider an isotropically scattering point S on the monochromator axis. We assume that the cell has a plane boundary of sufficient extent, and

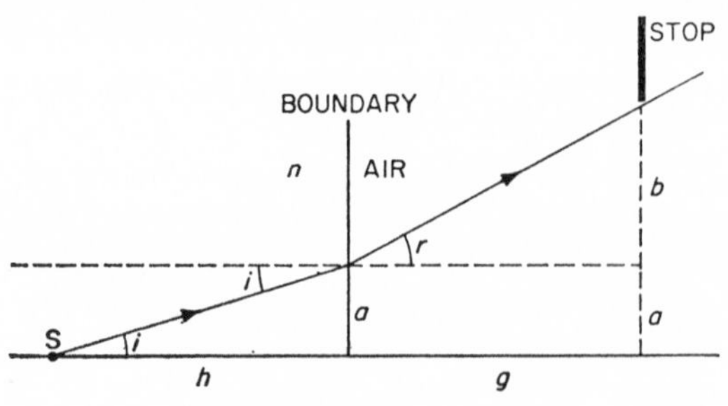

Figure 3.1

that the sample has refractive index n. Then, in all likelihood, the collection efficiency will be governed by a stop placed as shown. This might be a mechanical stop, or the image of the monochromator entrance slit or image slicer. Then:

$$a = h \tan i \tag{23}$$

$$b = g \tan r \tag{24}$$

$$n = \sin r / \sin i \quad \text{(more- to less-dense medium)} \tag{25}$$

$$a+b = \text{constant} = \text{physical stop radius} = k \tag{26}$$

If we consider small angles, then $\tan \alpha = \sin \alpha = \alpha$, and we find:

$$i = k/(h+ng) \tag{27}$$

Now, the collection efficiency for depolarized ($\rho = 1$) light or *small* stops depends, in the absence of large reflection losses at the interface, on the solid angle subtended by the stop. This is clearly proportional to i, whatever the shape of the stop, so we write the intensity:

$$I = k/(h+ng) \tag{28}$$

This relationship will clearly hold approximately for points off the axis, so long as the same stop is applicable to them. Thus, we may apply the equation immediately to systems with 90° illumination. The value of the parameters would vary from system to system, and g in particular might be used as a 'fudge factor'. Multipass systems would probably never be used for measuring relative intensities in different samples, owing to the difficulty in obtaining identical alignment, but in principle the expression:

$$I = \sum_m k_m/(h_m + ng) \tag{29}$$

would hold for m passes, where loss of laser beam intensity and collimation is included in k_m.

It is quite customary to use cylindrical cells, with a single pass of the laser down the cylindrical axis (see Fig. 2.2). Elementary considerations in that case lead to the expression:

$$I = k\sqrt{1/(h+ng)} \tag{30}$$

Although the Cary on-axis system would not be used with a multi-reflection cell for intensity work, it is possible to use a u.v. cell successfully, though with a 50-fold drop in signal/noise ratio. We may apply our treatment to that case with $n = n_1/n_2$ for the boundary between sample and cell wall, assuming that the cell has the same refractive index as the hemispherical lens [n_1 and n_2 are refractive index for sample and lens (silica) respectively]. It must also be assumed that there is an effective stop on the back surface of that lens, probably the image of the image slicer (not necessarily sharp at that point), i.e. of the 'sliced' monochromator entrance slit. This assumes only that the condensing lens collects everything transmitted by the hemispherical lens within that image. Then:

$$I = k\int_0^l 1/(h+ng)\,\mathrm{d}h = k\ln(1+l/ng) \qquad (31)$$

for a laser path-length l. The parameter g would vary with the cell wall thickness.

It must be emphasized that these ideas are purely speculative and have not been subjected to any experimental tests. We intend merely to illustrate a possible approach to laser systems.

3.1.3.1 *Reflection losses.* We have so far ignored reflection losses both at the laser beam entry point and at the Raman light exit from the cell. These losses also, in principle, vary with the refractive index of the sample. Cleanliness is clearly important.

3.1.3.2 *Larger apertures.* In the use of larger apertures, two approximations break down. The assumption that $\tan\alpha = \sin\alpha = \alpha$ is clearly no longer valid; however, the correct forms might be carried through with some loss of simplicity. The other, more serious, difficulty lies in the angular distribution of intensity radiated by an emitting dipole. It will be recalled that this is constant for all directions perpendicular to the dipole, but falls off with $\cos^2 i$ when i lies in the plane formed by the dipole and the required direction. Collection efficiency and the $\bar{R}^2/\gamma^2(R)$ ratio clearly alter in a most complex way for that reason.

Suppose we are collecting light propagating formally along z, where I represents the observed intensity which may or may not be analyzed for components I_x and I_y. Then, using the symbols previously defined, for an exciting vector E_z we have:

$$I = I_x + I_y = k(R_{xz}^2 + R_{yz}^2 + 2dR_{zz}^2)E_z^2 \qquad (32)$$

whilst for E_y

$$I = I_x + I_y = k(R_{xy}^2 + R_{yy}^2 + 2dR_{yz}^2)E_y^2 \qquad (33)$$

where d is a divergence factor. These equations express the fact that, for a finite collection angle, light originating from a radiating vector *along* the monochromator axis is partially collected. Clearly, this light ideally finds its way equally into I_x and I_y, and we define our 'total divergence' collection fraction as $2d$. This is acceptable so long as the 'stop' has at least four-fold symmetry. A stop of lower symmetry (e.g. a slit) requires d_x and d_y to be defined separately, where I_x and I_y are measured separately. This division is also affected by the position of the analyzer in the optical train.

Very little work has been done on the effect of a finite collection angle, apart from the general considerations equally applicable to arc excitation (see Hester[4]). The problem of *calculating* the divergence factor is quite difficult and requires numerical integration. Some work by Stenhouse[66] may be interpreted in this way, and it leads to the correlations of Table 3.2, between optical aperture [$\tan i = 1/(2 \times f$ no.), for $n = 1$] and d.

Table 3.2 Correlation between f number and divergence factor (after Stenhouse[66])

$f/$	1	2	3
$2d = d_x + d_y$	0·081	0·022	0·010

In systems where an analyzer is employed, there is the additional possibility that the component analyzed against may be transmitted when using a large aperture. Thus, for E_z:

$$\begin{aligned} I_x &= k[R_{xz}^2 + d_x R_{zz}^2 + \delta_x I_y]E_z^2 \\ &= k[R_{xz}^2 + d_x R_{zz}^2 + \delta_x(R_{yz}^2 + d_y R_{zz}^2)]E_z^2 \end{aligned} \tag{34}$$

The reader may verify that the δ-term exists, by viewing crossed polaroids in a skew direction. It is, however, removed by using a polaroid analyzer in the form of part of a cylinder, where y is the axis of the cylinder. This is of course possible only where y is spatially confined to a reasonable locus, so that the cylinder radius may be large compared with the uncertainty in position of y. Anyway, it appears that δ will be much smaller than d for a normal system.[67] In order to minimize the error δ, most instrument manufacturers place the analyzer after the collector lens in the spectrometer, where the beam is least convergent.

It is in principle possible to measure d for any real experimental set-up, and it then becomes a most useful empirical correction factor. This can be done, for example, by observing the depolarization ratio of bands known from theoretical considerations to be either depolarized [$\bar{R} = 0$] or fully polarized [$\gamma(R) = 0$]. The latter is, of course, not possible in a liquid, but for e.g. carbon tetrachloride $\rho_l < 10^{-2}$.[64] We consider the previous cases in order. Thus for case (ii)(a), using depolarized exciting light and 90° observation:

$$\rho_{obs} = \frac{I_z}{I_y} = \frac{6\gamma^2(R) + d_z[45\bar{R}^2 + 7\gamma^2(R)]}{45\bar{R}^2 + 7\gamma^2(R) + d_y[45\bar{R}^2 + 7\gamma^2(R)]} \tag{35}$$

For case (i)(b), using polarized exciting light and actually observing the depolarization of the Raman light emitted at 90°:

$$\rho_{obs} = \frac{I_z}{I_x} = \frac{3\gamma^2(R) + d_z[3\gamma^2(R)]}{45\bar{R}^2 + 4\gamma^2(R) + d_x[3\gamma^2(R)]} \tag{36}$$

The next method of obtaining a depolarization ratio, the usual practical method for 90° observations, turning the laser polarization, gives:

$$\rho_{obs} = \frac{I_{axial}}{I_{cross}} = \frac{I_y + I_z \text{ of (i)(a)}}{I_x + I_z \text{ of (i)(b)}}$$

$$= \frac{6\gamma^2(R) + 2d[45\bar{R}^2 + 4\gamma^2(R)]}{45\bar{R}^2 + 7\gamma^2(R) + 2d[3\gamma^2(R)]} \tag{37}$$

This was the arrangement envisaged by Stenhouse,[66] and this formula satisfactorily fits the results of his numerical integration (Table 3.2). Clearly, the ν_{obs} value for a fully polarized band gives $2d$ $(= d_x + d_y)$ directly in this instance. In the last possibility discussed for 90° observation, an analyzer rather than a polarization scrambler is used in front of the monochromator, in a similar arrangement to the previous one, giving:

$$\rho_{obs} = \frac{I_y \text{ of (i)(a)}}{I_x \text{ of (i)(b)}}$$

$$= \frac{3\gamma^2(R) + d_j[45\bar{R}^2 + 4\gamma^2(R)]}{45\bar{R}^2 + 4\gamma^2(R) + d_j[3\gamma^2(R)]} \tag{38}$$

where d_j is used for $d_x = d_y$, x and y being the same laboratory axis in this instance. See also p. 83.

For 180° observation, we have from case (iii):

$$\rho_{obs} = \frac{I_y}{I_x} = \frac{3\gamma^2(R) + d_j[3\gamma^2(R)]}{45\bar{R}^2 + 4\gamma^2(R) + d_k[3\gamma^2(R)]} \tag{39}$$

and in this case, if the *laser* polarization is turned, x and y are the same laboratory axis, so $j = k$. However, if the *analyzer* is turned (with subsequent use of a scrambler or application of a correction factor), $j \neq k$. For the case of $j = k$, a convenient form is:

$$\rho_{true} = \rho_{obs}/(1+d-d.\rho_{obs}) \tag{40}$$

and similar formulae may be worked out for the other cases.

When attempting to make accurate measurements with a configuration in which the laser polarization is turned, it is important to make correction for any change in the transmission of optical components, after and including the half-wave plate when the vector is rotated. Small changes may occur even if the half-wave plate is always left in the beam.

It is clear that changes in sample refractive index alter the divergence factor, since they alter the effective aperture of the collection system. It appears from the results of Stenhouse that this will only be important for apertures wider than about $f/2$.

On the Cary 81 180° system used with a non-internally reflecting cell, some preliminary experiments by Gilson and also by Gall[50] failed to detect any departure of ν_{obs} from $\frac{3}{4}$ for depolarized bands. Probably this system has a maximum aperture of about $f/1$ for samples at the hemispherical lens, so according to Stenhouse $d < 0{\cdot}04$. This would be very difficult to detect. However, there is evidence that the divergence factor is important in single-crystal observations,[68] for which it is found that $d \approx 0{\cdot}2$ when many crystals are stuck to the lens with Nujol or glycerol. It is not known to what extent this is accounted for by crystal imperfections and, more particularly, internal reflections in small crystals. It is interesting that, whilst internal reflections in a crystal certainly increase the divergence factor, they may have rather less effect on the state of polarization of the Raman light, particularly in a well formed crystal.

3.1.3.3 *Calibration.* It might seem that a correction could be made for most of the polarization and intensity errors mentioned, by use of a simple calibration procedure. This will often provide an approximate correction, but difficulties arise in accurate work.

Where an external standard is used, it is difficult to place the standard in exactly the same position as the sample. This is particularly unfortunate in multireflection systems. The same cell must, of course, be used for sample and standard. Even given geometrical accuracy, we have seen that the parameters vary with the refractive index of the sample. An external standard is quite useless with coloured samples. However, see the technique of Hendra and Qurashi for use with 180° illumination.[69]

Internal standards[59,60,70] are not altogether reliable because both the polarization and the intensity of the standard change according to environment. Although the posited 'internal field' part of this change (see Section 3.3.1) is hopefully thought to apply equally to all components, the result is always somewhat in doubt.

The one factor which may always be safely examined is the depolarization ratio of a depolarized band, since this ratio is independent of environment. Normally the sample will have such bands which may be used as internal standards. Thus the divergence factor can always be ascertained in systems free from other errors. It is also found that ν_1 of carbon tetrachloride (459 cm^{-1}) retains near-total polarization in most environments, and may be relied on in all except the most accurate determinations ($\rho < 0{\cdot}01$ in all cases measured[64]). Tetrahedral and octahedral ions are also moderately reliable in aqueous solutions. Probably the depolarization ratios of most standards are more reliable in different environments than are relative intensities. By using internal standards a fairly good depolarization ratio can normally be determined even with the least ideal collection systems. For systems subject only to a divergence error, the *d*-factor can similarly be determined by use of the formulae given. In this way meaningful relative intensities can also be obtained. We should note that meaningful data can always be obtained from completely polarized bands in samples whose refractive indices are not too different; e.g. see ref. 54. Reference 71 describes how useful results were obtained with internal standards.

3.2 Treatment for Powders

A finely divided powder will, almost invariably, completely depolarize all radiation passing through a very small thickness. There are simple macroscopic reasons for this, but even if they are removed by immersing the powder in a liquid of identical refractive index, the birefringence properties of most crystals still ensure depolarization. In principle, therefore, a cubic (and hence non-doubly refracting) crystalline powder immersed in a liquid of identical refractive index can be treated by the methods applicable to fluids. The depolarization ratio can be determined. Glasses can be similarly treated, but true glasses rarely give good Raman spectra.

It is not generally possible to observe a depolarization ratio in other powders. However, it is still of interest to assess what is observed in relative intensity measurements using powders. We assume not only that the exciting light and Raman light are completely depolarized by outside influences, but also that observed scatter comes from all possible directions of initial scatter and that the laser propagation direction is lost:

$$E = E_x = E_y = E_z \neq 0$$

So that, in terms of the averaged tensor:

$$I_x = I_y = I_z = k\,\overline{R_{ii}{}^2}\,E^2 + 2k\,\overline{R_{ij}{}^2}\,E^2 \tag{41}$$

Thus, defining a new constant k' we find

$$I_{obs} = k'E^2(\overline{R_{ii}{}^2} + 2\overline{R_{ij}{}^2}) = (k'E^2/45)[45\bar{R}^2 + 10\gamma^2(R)] \tag{42}$$

This result should hold for any scattering system. Some results obtained with the Cary on-axis system,[68] using individual tensor components obtained from single-crystal measurements, suggest that this ideal value may not be achieved in reality. The experimental results with the sodium dithionate system can be expressed:

$$I_{obs} = kE[45\bar{R}^2 + c\gamma^2(R)] \tag{43}$$

where $c \approx 7$, but the prediction is not very sensitive to the value of c for bands with high values of $\bar{R}$, i.e. the only reasonably intense features of the spectrum. Later work[50] with K_2SnCl_6 suggests $c \approx 9$.

3.3 Interpretation of Raman Intensity and Depolarization Measurements

The 'quasi-classical' treatment of Raman intensity, involving the interaction of 'classical' electromagnetic radiation with non-interacting, 'quantized' harmonic oscillators via a modulated polarizability, gives[3] for Stokes lines the intensity formula [cf. equation (22), Chapter 1]:

$$I_i \propto \frac{(\nu - \nu_i)^4}{\mu_i \nu_i (1 - e^{-h\nu_i/kT})} \tag{44}$$

where I_i is the scattered intensity, per unit solid angle, per mole, per unit illuminating intensity; μ_i is the reduced mass of the harmonic oscillator, ν_i its frequency, and ν that of the excitation. For anti-Stokes lines the formula is:

$$I_i \propto \frac{(\nu + \nu_i)^4 e^{-h\nu_i/kT}}{\mu_i \nu_i} \tag{45}$$

It is possible to show that the formulae also apply to the *integrated* intensity expected for vibration–rotation bands of rotating harmonic oscillators. Consequently these formulae are normally used as a basis for the interpretation of Raman intensity in colourless liquids and gases.

We now summarize some of the more purely quantum theoretical approaches to the problem, and some refinements to the quasi-classical formulae, in part justified by them. It was perhaps insufficiently emphasized in Chapter 1 that Raman scattering is a second-order process. In simple terms this means that approximations which are sufficient to account for resonant processes fail to predict the phenomenon at all. The difficulty in this situation is to decide which of the many approximations should not have been made.

A paper by Ting[72] surveys some of the available treatments and suggests a quantum theoretical approach which is stated to be generally valid. The authors are not competent to judge the acceptability of this and other recent treatments, which have not yet attracted comment from the scientific community, but it is to be hoped that something useful will emerge from them.

In the quantum theoretical treatment, information about Raman intensities comes eventually from various 'sum rules' for Franck–Condon integrals. According to Ting, the general quantum theoretical light-scattering expression can be manipulated in such a way that a previously ignored factor is shown to result in Raman intensity, by use of an unusual formulation for the sum rules. The result is that Raman scattering is allowed in second order, in the presence of rigorous vibrational, electronic, and rotational wave-function separation (the Born–Oppenheimer approximation), but only for totally symmetric bands and diagonal Raman tensor elements. In a previous treatment due to Albrecht[73] it was suggested that all Raman intensity arises from the breakdown of the Born–Oppenheimer approximation (particularly by vibronic coupling). Ting therefore suggests that 'Born–Oppenheimer allowed' terms are responsible for the more obvious features of Raman spectra, but that weaker features appear, due to the breakdown of rigorous separability. It is certainly qualitatively correct that, in general, totally symmetric bands give rise to the strongest features of Raman spectra.

It is clearly difficult to assess the relative importance of vibronic coupling and Born–Oppenheimer allowed terms in totally symmetric transitions on the Ting formulation, in the liquid state. Ting suggests that vibronic coupling becomes highly important only when the molecule has low-energy electronic excited states, to which the principal transition-moment contribution arises from vibronic coupling with a high-energy state to which the transition is strongly allowed. In particular this includes nearly all aromatic compounds.

It seems clear that, whatever its limitations, the Albrecht theory contains the key to those anomalies in liquid Raman intensities which are directly attributable to ground-state intermolecular interactions. Thus, Wall and Hornig[63] point out that a dependence on refractive index (expected classically) is implicit in the Albrecht treatment, and they go on to consider the case of binary solutions, and vibronic mixing between solute–solvent wave functions. With such mixing considered, it is predicted, in line with some (but not all) observations, that, for the component of a binary solution with the lowest energy electronic transition, intensity is lost equally from all the bands to all the bands in the other component. This treatment contains the dependence on refractive index in implicit form, but it is more general than the 'macroscopic' theory (see later). At the same time, deviations still occur, hopefully corresponding to mixing of the ground-state wave functions or, in other words, specific local interactions.

3.3.1 *Classical Procedures*

Despite the careful attention now being given to quantum theoretical procedures, it remains true that most of the useful predictions and methods are classical or semi-classical; however, see Krushinskii and Shorygin,[74] Rebane,[75] and Shorygin and Ivanovna.[5]

The quantum formulation of neighbouring-molecule interference with the scattering process[63] is one of the only practically useful results to appear from the strict quantum approach. However, the classical or macroscopic treatment is more definite in its predictions (even if less valid). The macroscopic approach derives from the macroscopic form of the Maxwell equations (see Chapter 1) which consider a uniform medium of refractive index n, where n is simply related to the optical frequency dielectric constant. A scatterer embedded in this medium interacts in a damped way with the electric field of the exciting radiation, and the induced dipole similarly radiates less vigorously. If the variation of the refractive index over the Raman shift is ignored, the reduced intensity is given[76] by:

$$I_r = 81 I_0/(n^2+2)^4$$

The resulting correction is known as a 'local field' or 'internal field' correction. See, for example, Tulub,[77] and note also Fini *et al.*[78] It should be applied whenever an attempt is made to interpret intensities in different samples in a classical way. For example, in the work of Bahnick and Person,[71] stable donor–acceptor complexes are suggested between carbon tetrachloride and various solvents.

3.3.2 *The Wolkenstein Theory*

This classical theory[79] attempts to derive the various $\partial\alpha_{\rho\sigma}/\partial q_i$ values on the assumption that only *bond* polarizabilities change during a vibration, and that only bond-stretching effects this change directly. Deformations are supposed to acquire activity only by virtue of the change of aspect of the bonds, or by mixing with stretches. However, Woodward and Taylor have suggested a simple modification to permit direct changes. Sverdlov[80] has also discussed the general 'first order' Wolkenstein approximation, and its application to overtone and combination bands.[81]

Woodward and Long[54] further attempted to obtain correlation between bond type and parameters obtained by using this theory. They find that for certain XY_4 molecules the empirical relationship:

$$\overline{\partial\alpha/\partial r} = Cp(Z_X+Z_Y)$$

holds, where r is the internal bond-stretching coordinate, Z_X and Z_Y are the atomic numbers of X and Y, and

$$p = 100\exp[-\tfrac{1}{4}(X_X-X_Y)^2]$$

where X_X and X_Y are Pauling electronegativities. Of particular practical importance is the prediction of very low activity for largely ionic bonds, a conclusion fully borne out by experiment. This is particularly evident for largely ionic solids which, even when they have Raman-active fundamentals, are often very poor scatterers.

The Wolkenstein theory, its developments, and its use with the Wilson F-G matrix technique, are discussed by Hester.[4] Developments in this field are summarized about once every two years in *Analytical Chemistry*, and the 1966 review[82] is a particularly useful source of references to Russian work.

The Wolkenstein theory was used recently (with a very approximate assumption for the form of the normal vibrations) in an attempt to relate the laser-excited Raman intensities of the two totally symmetric modes of $M(CO)_6$.[69] It appears from this treatment that $\partial\alpha/\partial r$ for the stretching of the metal–carbon bond decreases in the order Re > Mo > W > Cr > V, and this was correlated with decreasing π-bonding. In the sodium dithionate single-crystal experiment[68] it was found that not only the solution depolarization ratios but also the solution relative intensities were very close to those predicted using equations (1)—(4), (11), and (12) and the single-crystal Raman tensor elements. The Wolkenstein procedure was also applied to the $S_2O_6^{2-}$ ion, using a force field transferred from ClO_4^-, and reasonable consistency was obtained.

We should now consider the quantum theoretical restrictions which should be placed on applications of classical theory. Unfortunately little appears to have been written explicitly on this subject since Placzek enumerated the basic conditions, namely, a non-degenerate electronic ground state (but see Child and Longuet-Higgins[83]), and excitation much higher in frequency than the vibrational fundamental frequencies, but much lower than the lowest electronic transition frequency. One Russian paper may be relevant.[74]

If Ting is correct in suggesting that two different mechanisms account for Raman activity, it is not immediately apparent that a meaningful classical comparison can be made of intensity observed for, particularly, totally symmetric and non-totally symmetric bands. However, a detailed analysis might reverse this judgment. It is clearly true that a great deal of caution should be exercised where the refractive index varies, in case the macroscopic approach (local field correction) is not strictly valid.

3.3.3 *Band Contours*

The band contours of vibrational spectra for gaseous samples, when not resolved into discrete bands, can still be used to determine band type; see e.g. ref. 2. In principle, liquid band contours also contain information on molecular rotation and hence, indirectly, on molecular interaction. Gordon[84] suggests the use of the Fourier transforms of the spectra in order to examine this effect. An application of band contour analysis is summarized in Chapter 6.1.

A feature of Raman rotation–vibration spectra is the occurrence of a narrow Q-branch of much greater intensity than other branches, for totally symmetric vibrations. Non-totally symmetric vibrations have very weak Q branches, but prominent O, P, R, and S branches (see Chapter 5). These generalizations are undoubtedly associated with the observed narrowness of totally symmetric bands in liquid-phase spectra.

[Murphy, Holzer, and Bernstein (N.R.C., Ottawa, Canada), at the International Raman Conference at Carleton University, Ottawa, in August 1969, largely confirmed the accuracy of equations (36) and (37) and Table 3.2.]

4

Single-crystal Raman spectroscopy

4.1 Introduction

It was observed in Chapters 1—3 that the advantage of single-crystal spectroscopy lies in the fixed orientation of all parts of the crystal with respect to each other and to the spectrometer. It was also noted that in molecular crystals these advantages are offset to some extent by the crystal field. In this chapter all these ideas are put on a formal basis. We shall also see that, in any crystal, the (first-order) Raman tensor elements can be classified according to the crystal 'point symmetry'. This, in centrosymmetric crystals, gives rise to rigid selection rules. We recall also that a Raman tensor element R_{ij} is involved in scattering in which the exciting electric vector lies along j, and the scattered light appears as if from a generating dipole along i. Normally $R_{ij} = R_{ji}$. In this chapter i and j will generally be 'crystal axes'. The importance of laser excitation lies in the ease with which one direction of the exciting vector is selected.

The size and quality of crystal which can be examined depends largely on the precision required of the measurements. It is possible to assign crystal modes to crystal point symmetry with crystals having smallest dimension down to fractions of a millimetre. The imperfection of the measurements arises from internal reflections in otherwise perfect crystals, and from this point of view the spectrum of an absorbing crystal, if it can be obtained, may be more ideal in terms of symmetry. In general the absorption reduces the value of measured intensities.

It is possible to apply the 'oriented gas phase' approximation to molecular crystals. To do this, somewhat more accurate intensity measurements are required, and it is preferable for the minimum dimension of the crystal to be at least a few millimetres in such a case.

More elaborate measurements require even larger crystals, perhaps

1 centimetre in diameter. In principle some of the internal reflection troubles could be avoided by immersing the crystal in a medium having a similar refractive index. This is not a very convenient method for routine use, and anyway some crystals have much too high a refractive index.

Although there are important differences in the infrared and Raman spectroscopy of crystalline materials, a single theory of the vibrational properties of crystals is basic to both. We shall start by summarizing this theory. This is also the natural place to discuss polymers, since a crystal is effectively an 'infinite' three-dimensional polymer. In this context we define an 'infinite polymer' as one so great in extent that its vibrational properties are independent of size. The vibrational analysis of such materials is outside the scope of this book.

4.2 Vibrations and Symmetry

The reader should be aware of the way in which the vibrations of molecules depend upon the symmetry of the molecule. In general a vibrational mode has a specific effect on each symmetry operation of the molecule; some operations are retained and some are lost. It is possible to classify the modes according to the 'symmetry types' of the point group in this way, or to assign normal coordinates as bases for the 'irreducible representations' of the point group in more technical language. It is also possible to conceive of 'symmetry coordinates', which describe a possible idealized vibration of the molecule, and belong to one of the symmetry types of the point group. An actual vibrational mode 'normal coordinate' consists of a mixture or a linear combination of all the symmetry coordinates belonging to the *same symmetry type* only.

In polymers and crystals we have to consider new types of *translational* symmetry element. For these to be perfect, the crystal would clearly have to be of infinite extent. However, the number of units in any tractable piece of crystal is clearly very large indeed, and we find that no discrepancies arise due to the introduction of infinite size as a mathematical fiction. For polymers of 'intermediate' molecular weight we may need to make allowance for the length effects.

These new symmetry elements assume a very similar importance to the familiar non-translational point group elements. However, a certain amount of unfamiliar jargon used by physicists and mathematicians is attached to their use. We discuss crystal symmetry briefly, and then proceed to show the relationship between crystal symmetry and crystal vibrations.

Before that, however, we wish to formalize this qualitative discussion of the symmetry properties of vibrations, in a way which is sufficiently general to be applicable to the 'primitive translation' operation which we discuss later. We have first to define a 'group' of symmetry operations in the mathematical sense. In mathematics the word 'group' is not used in an unrestricted way; the general-purpose word with equivalent lay usage is 'set'. For a set of operations to constitute a group it must have certain properties.[85,86] The most important are that the identity operation should be a member of the set, and that the successive application of two operations should be equivalent to some other operation of the set.

We may pick on any atom of an assemblage, and apply to it the group of symmetry operations applicable to that assemblage. Some of these operations move the atom from its original position, and some do not. We call the sub-group of operations which does not move it the 'local' or 'site' group for the atom concerned.

We consider now one operation R^1 which does move the atom into the position of another, symmetrically equivalent, atom. By successive operations R^1, where the overall group is *finite*, the atom is eventually moved back to its original position. If this occurs after p applications, we write:

$$R^0 \equiv R^p \equiv E \tag{1}$$

the identity operation. The atoms encountered during this process constitute a symmetrically equivalent set, though not necessarily a full set. If not, it is termed a 'sub-set'. Where the overall group is *infinite*, there may or may not be a value of p such that $R^p \equiv E$ for any particular R, but the atoms encountered are always a symmetrically equivalent set or sub-set, of infinite extent where p does not exist.

The symmetry of the assemblage of atoms is, as already indicated, an effective constraint on the forms of its normal modes of vibration. The constraint on a particular atom is quite analogous to that on the assemblage: it may move only along coordinates which form a basis for the irreducible representations of its sub-group; that is, the coordinates must be classifiable according to the symmetry species of the sub-group. The only partial exception to this rule is in the case of multidimensional irreducible representations of the overall group. It is always possible to take components of degenerate modes that appear in these circumstances, in such a way that the rule is satisfied for a given symmetrically equivalent set of atoms. However, it may be necessary to take different components to satisfy the rule for a different set. We consider later the 'correlation theorem' which relates sub-group and overall group symmetry species.

We wish here to draw attention particularly to the p allowed combinations of the set of p equivalent coordinates, generated from one atomic coordinate by successive application of R^1. We define a parameter ϕ, named, for reasons which will be apparent, the 'space phase' difference allowable for the set of p coordinates, in R-space. This parameter takes the possible values:

$$\phi = 0, 2\pi/p, 4\pi/p, \ldots \quad \pi \quad (p \text{ even}) \tag{2}$$

$$\text{or} \quad \ldots (p-1)\pi/p \quad (p \text{ odd}) \tag{3}$$

Two allowed linear combinations of the p coordinates X_r correspond to a particular value of ϕ, and are, for the purist:

$$N_1 \sum_{r=0}^{p-1} e^{ir\phi} X_r \tag{4}$$

and its complex conjugate

$$N_2 \sum_{r=0}^{p-1} e^{-ir\phi} X_r \tag{5}$$

where the N_n are normalizing factors. The realist, however, prefers:

$$N_3 \sum_{r=0}^{p-1} X_r \cos(r\phi) \tag{6}$$

$$N_4 \sum_{r=0}^{p-1} X_r \sin(r\phi) \tag{7}$$

These are obtainable from the former pair (4) and (5) by virtue of the standard relationships:

$$\cos\theta = \tfrac{1}{2}(e^{i\theta} + e^{-i\theta}) \tag{8}$$

$$\sin\theta = (1/2i)(e^{i\theta} - e^{-i\theta}) \tag{9}$$

We note that for $\phi = 0$ or π, $e^{ir\phi} = e^{-ir\phi}$, so that the 'imaginary' pair are not different allowed combinations in these two cases (nor are they in fact imaginary). Similarly the sine function is zero for these values of ϕ, and may be discarded. For other values of ϕ, each pair is degenerate, which is the justification for accepting apparently arbitrary mixtures within the pair.

These linear combinations are a form of symmetry coordinate. Each individual or pair can be assigned to a symmetry species of the overall group. We show later, where it is more relevant, the way in which this is done, but meanwhile we note that these ideas apply to sub-assemblages of atoms as well as to individual atoms. For example, sub-set allowed combinations may be built up into full-set allowed combinations in this way.

4.2.1 *Crystal Symmetry*

This book would not be a suitable place for a detailed exposition of crystal symmetry, since this appears in a number of standard texts. We summarize the ideas required for vibrational work, so that the reader will know what knowledge is involved.

4.2.1.1 *The primitive cell.* We define a crystal (with hindsight) as a regular array of atoms suffering attractive and repulsive forces. A certain 'unit cell' is chosen to describe the structure. This cell is usually defined by three directions (axes) not necessarily at right-angles, three corresponding lengths, and a point of origin. The cell is normally thus a parallelepiped; the crystal is composed of a stack of identical cells.

Certain conventions have been established by crystallographers for the choice of unit cell. This is quite arbitrary, and it is important for us to go into the nature of the choice, and the alternatives. Not all unit cells have the same volume, but a certain smallest possible volume exists, and any cell having this volume is called a 'primitive cell'. It is most important for vibrational purposes to know the number of primitive-cell volumes contained in whatever conventional cell is being used. In the 1952 edition of *International Tables for X-Ray Crystallography* (see bibliography at end of this chapter) the space-group symbol unambiguously indicates the nature of the conventional cell, as shown in Table 4.1. These 'centred' cells are used because they have the full symmetry of the lattice.

Table 4.1 Conventional Unit Cells

Symbol	Number of primitive cells
P	1 (Primitive)
R	1 (Rhombohedral)
A	2 (Centred on '*A*' faces)
B	2 (Centred on '*B*' faces)
C	2 (Centred on '*C*' faces)
I	2 (Body centred)
F	4 (Centred on all faces)

The three vectors defining a primitive unit cell also define the symmetry operation known as a 'primitive translation'. Clearly this is the operation relating an atom in one primitive cell to the identical atom in a neighbouring cell. The choice of primitive cell and origin is arbitrary but governed by conventions; a number of cells have the same smallest possible volume (see Figure 4.1). However, it is strictly sufficient to make a choice and to keep to it.

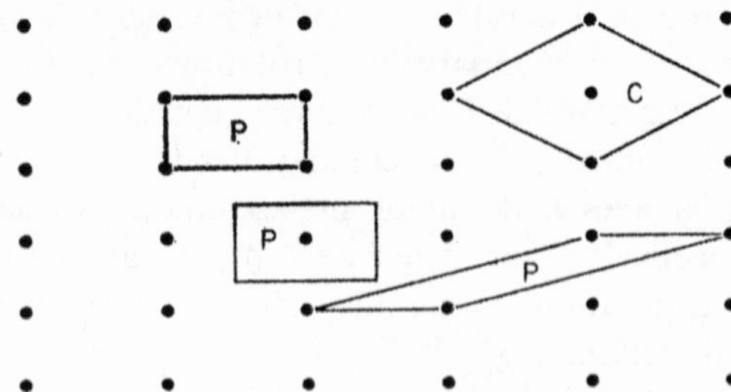

Figure 4.1. Two-dimensional illustration of different choices of primitive unit cell (P) and a centred cell (C)

It may be helpful for vibrational work to make a number of unconventional modifications. The origin may be moved so that the cell faces do not intersect atoms. It is not necessary that the primitive cell should be a parallelepiped. Other stacking bodies are defined by the use of more than three crystal axes. The hexagonal and triangular section right prisms are occasionally encountered. Another example, generated by four equi-inclined axes, is the rhombic dodecahedron. In this way it may be possible to choose a primitive cell with full symmetry. The reader should not feel obliged to do this; it is mentioned as a possible help in thinking round a difficult problem.

4.2.1.2 *Space group and factor group.* The 'space group' of a crystal is simply the group of all symmetry operations applicable to it. To make this a true group with perfect symmetry operations, we have so far considered the 'infinite crystal' approximation, for which the space group is of course an *infinite* group. For some purposes it is more helpful to have a *finite* group. For this purpose we introduce, effectively, a 'finite space' in which a traveller proceeding in a straight line inevitably returns to his starting point. It is possible to retain mathematical consistency within this 'curved space' with its non-Euclidean geometry. Indeed, if we accept the hypothesis of general relativity, that this is the true nature of space, we have merely contracted its 'radius' for a special purpose.

The justification for accepting either the infinite or the cyclic approximation is simply that we know experimentally that some properties of crystals are independent of size. It is then reasonable to suppose that these *bulk* properties will be adequately explained using such approximate models.

All the normal point-group operations (rotation axes, mirror planes, rotation–reflection axes*) may be present in a crystal. Crystallographers usually use, as the third necessary type, rotation–inversion axes, but we shall see that it is preferable to retain the (equally valid) point-group convention for vibrational work. There may also be translational generalizations of rotation axes and mirror planes, known as screw-axes and glide-planes respectively. The screw-axis M_n is a rotation of $2\pi/M$, together with a translation, along the axis, of n/M times the repeat interval in that direction.

Glide planes consist of a reflection, together with a translation in the plane of reflection, of half the repeat interval in the translation direction. Symbols *a*, *b*, *c*, *n*, and *d* are used according to the relationship between the translation direction and the conventional cell edges; see *International Tables*.

In an infinite crystal, every symmetry element occurs an infinite number of times. Symmetry elements related by primitive translations clearly have closely related properties. Also, there are some glide planes and screw axes (occurring with non-primitive cells) which differ in effect from mirror planes and rotation axes in other positions only by one or a succession of primitive translations. These we may often discuss together with their 'parent' non-translational operations.

Some crystal phenomena do not affect the operation of primitive translation. This will be the case where every primitive cell is similarly affected; we would be interested in a vibration where the movement in every primitive cell is identical. We are justified in ignoring primitive translations in the symmetry classification of such phenomena. The group of symmetry operations which remains, after primitive translations have everywhere been equated to the identity operation, is called the 'factor group'. We would expect phenomena having only a long-distance deviation from total uniformity in every cell to be classifiable according to the factor group symmetry.

Every possible factor group is isomorphous with a point group, meaning that the kind and number of operations is the same, but that some of the point group operations are replaced, in the factor group, by their translational equivalents: rotation by screw axes, and mirror by glide planes. The isomorphous point group ('point symmetry') is given for all space groups in crystallographic reference works. It is also indicated by the Schoenflies space group symbol, which is merely the point symmetry with a serial superscript running through the possible substitutions of non-translational by translational operations. The irreducible representations of a factor

*Note that $i \equiv S_2$, so this is a complete list.

group are the same as those of its isomorphous point group, and there are great similarities in its use. We consider factor group analysis later.

4.2.1.3 *Plane and line groups.* The ideas of the previous section are easily extended to linear and planar polymers, with, respectively, primitive translations in one or two directions only. Conventionally, the group of all symmetry operations applicable to such a polymer is called a line group in three dimensions, or a plane group in three dimensions. The number of dimensions needs to be specified to distinguish, for instance, the (two-dimensional) projection of a space group, commonly called simply a plane group. Clearly the difference lies in permitting symmetry elements to appear in the third dimension, where they have no translation in that dimension. This nomenclature is too cumbersome for general use, and in this book we refer to the three-dimensional groups with initial capital letters on all occasions: Space-group; Plane-group; Line-group. The reader must appreciate that three dimensions are understood when these terms are encountered. Should we need to refer to a reduced number of dimensions, we would then use the lower case initial letter.

These groups will be encountered with reference not only to autonomous polymers but also to crystalline materials having internal groupings 'infinite' in only one or two directions. We define the 'site group' later, but we note now that all Plane-groups, and Line-groups which are site groups arising in crystalline materials, are identical with one of the 230 possible Space-groups, the only restriction being that on the number of dimensions in which the primitive translation operates. There are 80 possible Plane-groups and 75 crystallographically restricted Line-groups.

The Line-group of a 'free' linear polymer is not subject to the restrictions imposed by the filling of space or plane, and in principle there is an unlimited number of them. This is because such a Line-group may be formed with any n-fold rotation axis.

4.2.2 *Other Types of Array; Helical Polymers*

Materials which do not form polymers, or which do so only on solidification, almost invariably take up a structure which, if regular at all, has crystalline regularity as discussed in the previous sections. It is possible to conceive of limitless other types of regularity, in which primitive translations are not present (or involve an inconveniently large cell) but in which certain other symmetry elements are present. A simple and important example is the helical polymer, with its only symmetry element a screw axis other than one of the crystallographically acceptable ones.

In crystals the element M_n has $M = 1,2,3,4,6$ only, and $n = 1$ to M only. No other possibility can exist in conjunction with a primitive translation. (Without such a translation the operation must, of course, be re-defined in terms of absolute lengths and angles.) Such helical polymers are well characterized materials; some proteins and synthetic polypeptides take this form. Higgs[85] has pointed out that the group of all operations for such a molecule is isomorphous with the point group C_{∞}, and classification may be made accordingly. This method might be generalized if other cases are encountered. It is also useful when it results in a repeat unit which is much smaller than that defined by the primitive translation.

4.2.3 *Vibrations in Crystals*

Vibrations in crystals have been studied mathematically in the usual way, by setting up equations of motion for the atoms in simple model systems, and searching for periodic solutions. In this context the mathematical fiction of an infinite crystal is necessarily abandoned, and the physical fiction of cyclic boundary conditions is substituted. These conditions, which are closely related to the finite space model, assume that the atoms on one face of a crystal are rigidly linked to the identical atoms on the opposite face. This greatly simplifies the mathematics but would again not be expected to affect solutions describing bulk properties of crystals.

We consider a crystal with N primitive unit cells, each containing n atoms. Clearly there is a total of $3nN$ degrees of freedom arising from motion of the atoms, and in principle we expect $(3nN-6)$ non-zero solutions to the vibrational problem. This astronomical number has to be dealt with by classification rather than enumeration of the modes.

The solutions to this problem bear an important relationship to other standing-wave phenomena. For instance, it is useful to picture the modes of a violin string. For crystals, in one vibrational mode all the particles in the crystal oscillate 'in phase' and therefore at the same frequency. By 'in phase' in this context we mean that the time dependence is the same throughout the crystal. As with other standing waves, the 'amplitude' of vibration varies in space through the crystal, and unfortunately the periodic properties of this fluctuation are described by a 'space phase' difference between neighbouring primitive cells.

Our previous discussion of allowed space-phase differences was limited to closed or finite spaces and groups, and the analysis of crystal properties must always take account of the finite extent of the crystal in some way. To apply symmetry arguments to the overall crystal, we must clearly use the 'finite space' model. In this purely conceptual model of a crystal, when a crystal boundary is crossed in the course of a symmetry operation, it is assumed that the opposite face is entered in the corresponding position. In this way it is possible to consider the allowed space phase difference in primitive-translation space; this parameter is referred to as the mode 'wave vector' and is given as:

$$\boldsymbol{k} = 2\pi/\boldsymbol{\lambda} \tag{10}$$

where $\boldsymbol{\lambda}$ is the wavelength in units equal to a primitive translation. This 'natural' unit of length is convenient for many crystal phenomena, though it is not used universally. It may be seen that, since the crystal is not ultimately homogeneous, values of $|\boldsymbol{\lambda}|$ less than 2 have no obvious definition. For reasons of mathematical consistency, for $|\boldsymbol{k}| > \pi$ we define the wave vector as:

$$\boldsymbol{k}' = 2\pi - \boldsymbol{k} \tag{11}$$

so that solutions with $|\boldsymbol{k}| > \pi$ are simply identical with other solutions at $|\boldsymbol{k}| < \pi$. The region of $|\boldsymbol{k}|$-space $0 \leqslant |\boldsymbol{k}| \leqslant \pi$ is called the '(first) Brillouin zone'; a wave with $|\boldsymbol{k}| = 0$ is 'at the zone centre' and with $|\boldsymbol{k}| = \pi$ it is 'at the zone boundary'.

Referring now to our discussion of space and factor groups, we see that crystal vibrations at or near the zone centre, and only such vibrations, may be classified according to the factor group. It turns out that infrared absorption and Raman scattering occur in first order only by virtue of interaction of photons and crystal vibrations of similar wavelength,* and such vibrations are close to the zone centre; hence, the vibrations observed in this way may be classified according to the factor group symmetry types. The selection rules for them are the same as for classifications under the isomorphous point group.

This important result sometimes obscures the rest of the crystal vibrations. The observed zone-centre modes are called factor group fundamentals or, by Mathieu, 'oscillations principales'. Vibrational fundamentals of the crystal occur all through the zone, and they are all equally responsible for phenomena such as specific heat; they may in principle be observed by inelastic neutron scattering. Quanta of energy in crystal modes are referred to as 'phonons'.

We have already observed that a crystal has $3nN$ degrees of freedom, and $(3nN-6)$ vibrational modes, i.e. effectively $3nN$ modes since N is very large. At the zone centre, where the factor group applies and primitive

*This is not true of forward scattering.

translations are treated as identity operations, we are clearly only considering the $3n$ degrees of freedom arising from movements of the atoms present in one primitive cell. With this in mind, we may consider crystal vibrations in a way which starts with these $3n$ degrees and generalizes to include all unit cells. This treatment is confined to crystals without translational symmetry elements other than primitive translations, and which are capable of being described with a unit cell of full symmetry which, however, contains no split atoms. The results are nevertheless applicable to all crystals.

Consider first N non-interacting groups of n atoms, each group of n being the contents of one primitive cell of the type defined. Each cell has $(3n-6)$ vibrations, 3 translations, and 3 rotations. The $(3n-6)$ vibrations are classified according to the factor group of the crystal, which in this case is not only isomorphus with a point group but identical with it. If we now introduce coupling between the cells, we see that the result must be that each of the $3n$ motions of one cell couples in N different ways to give N overall modes. We have already seen that for such of the $3nN$ as are close to the zone centre, having an approximately equal amplitude in all N cells, we may retain the factor group classification. Of course one 'factor group fundamental' as spectroscopically observed is really the envelope of a large number of crystal modes falling within the definition.

It may be seen that both vibrations and rotations of the unit cell give rise to crystal vibrations at all positions in the Brillouin zone. For that reason there are $(3n-3)$ factor group fundamentals, and pure unit cell rotations may not be factored out as they are in molecular vibrations. For Line-groups there are correspondingly $(3n-4)$ factor group fundamentals. Crystal vibrations arising from these $(3n-3)$ primitive cell vibrations are known as 'optical modes'. At the exact zone centre the three translations of the unit cell clearly become the three translations of the crystal; away from the centre, however, they give rise to crystal modes known as 'acoustic modes' since they are essentially standing sound waves present in what is, not too far from the zone centre, a homogeneous medium. These acoustic modes may also give rise to Raman scattering (Brillouin scattering). It is usual to *define* acoustic modes as those which have zero frequency in the zone centre version; they have a very small frequency at the wavelengths appropriate to Raman scattering. It is these acoustic vibrations which largely account for the specific heat of crystalline solids at and below room temperature.

The problem of defining the N allowed combinations of each unit cell mode is easy only in the case where coupling between the cells is minimal. In that case the unit cell modes retain their identity away from the zone centre, and starting with one mode of one cell we define the allowed combinations in the direction of one primitive translation as:

$$\sum_{r=0}^{N^{\frac{1}{3}}} q_r \cos(r|\boldsymbol{k}|) \tag{12}$$

$$\sum_{r=0}^{N^{\frac{1}{3}}} q_r \sin(r|\boldsymbol{k}|) \tag{13}$$

where $|\boldsymbol{k}|$ takes the values $0, 2\pi/N^{\frac{1}{3}}, 4\pi/N^{\frac{1}{3}}, \ldots$, and q_r is the normal coordinate for the selected cell mode in cell r. Every allowed combination within this line of cells, defined by this one direction of translation, combines within the plane now formed from it by primitive translation in a second direction, in the same way. This plane similarly goes into the whole crystal on application of the third primitive translation. Thus, one cell mode becomes $(N^{\frac{1}{3}})^3 = N$ modes in the crystal as required. Unfortunately this simple description is not adequate for real crystals, except where the wave vector is small. Outside the range of applicability of the factor group, the unit cell modes are not such a useful starting point since their form and frequency vary according to the value of the wave vector. Although the problem does not immediately concern us in this book, we pursue it a little further to show why it is that discussion of second-order effects is so limited.

Since N is very large, the allowed (space) phase difference between cells is for practical purposes a continuous function. We consider one observed zone-centre mode. This is really a large number of crystal modes, with a quasi-continuous range of wave vector values moving gradually away from zero. As the departure from the zone centre becomes serious, two things happen. The frequencies of all the crystal modes change from the zone-centre value (this is usually described as dispersion), and the form of the mode within one cell also changes. Where several factor group fundamentals of the same symmetry occur, the changes allowed by factor group symmetry may occur first. More serious than this, as the factor group classification breaks down, other levels may be allowed to interact and to mix. We may imagine that, owing to the dispersion, some levels approach one another; a 'non-crossing rule' may be invoked if the interaction of the levels is at all allowed, and the result is that, away from the zone centre, ordinary factor group symmetry considerations tell us nothing about the crystal vibrations. This discussion tells us nothing about the extent of the interaction of levels, which may in some cases be very small. In such a case the non-crossing rule takes the form shown in Figure 4.2, which is not practically different from crossing, away from the critical region.

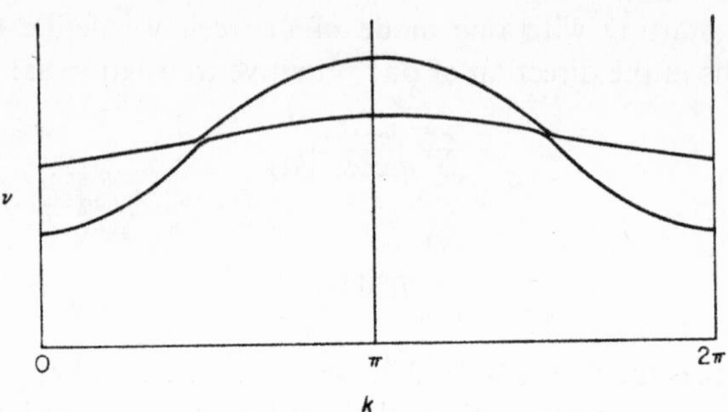

Figure 4.2 Form taken by non-crossing levels, for very small interaction

In second-order infrared and Raman spectra, the restriction to zone-centre modes is no longer present. As might then be expected, the factor group symmetry restrictions on the activity of such processes also break down. Observed second-order intensity changes only slowly with wave vector; the result is that second-order crystal spectra are often extremely broad, owing to the dispersion.

Although the simple form of the factor group classification breaks down away from the zone centre, the symmetry constraints imposed by the space group are still rigorously obeyed. This includes, in a formal way, the constraint upon atoms in a particular set of sites. The difficulty is that an infinite group has an infinite number of irreducible representations, and some of these are, in general, multidimensional. Similar difficulties appear in the cyclic model, where infinity is replaced by a very large number. The atomic site approach is valid only for a particular choice of components of degenerate crystal modes, and in general for a different choice for different sets of atoms. Clearly this method is not viable in such circumstances.

We consider later, in connection with real supercells, the second-order supercell approach. Here we note again that the astronomical number of crystal modes can be dealt with only by classification. A particularly hopeful approach considers certain critical wave-vector values, and the smooth transition of properties between such values.[86] For some purposes the symmetry properties of crystal modes are of less interest than the number of them occurring within a particular frequency increment. For example, specific heat depends upon this. This alternative way of presenting the results may be encountered.

Where factor group fundamentals are used as a basis for a separate set of lines on either a dispersion curve (frequency against k) or a frequency increment curve (frequency against number of modes in a given frequency increment), these lines are customarily referred to as the appropriate 'optical branches' and 'acoustic branches'.

4.2.3.1 *Transverse and longitudinal modes.* Although the general problem of crystal vibrations has not been solved in three dimensions, we may say a little more about the expected nature of the solutions. Considerations of symmetry lead us to expect a defined 'phonon propagation direction'. This does not mean that the phonon (a standing wave) is actually travelling, but that it has a wave vector with a defined direction throughout the crystal; the direction associated with an observed phonon depends on the nature of the observation. In our previous formulation, we considered the value of the wave vector in three chosen 'primitive translation' directions. These may be considered as the three components of the overall wave vector, which is therefore obtained as their vector sum. For Raman scattering there is a simple rule stating that the wave vector is conserved or:

$$\boldsymbol{k}_i = \boldsymbol{k}_s + \boldsymbol{k} \tag{14}$$

where $\boldsymbol{k}_i$ is the wave vector of the incident light, $\boldsymbol{k}_s$ that of the scattered or Raman light, and $\boldsymbol{k}$ that of the excited or Stokes phonon. For this purpose we must use absolute length units, to avoid a change in units with direction. We see that, for all except forward scattering, $\boldsymbol{k}$ and $\boldsymbol{k}_i$ are similar, as stated previously, and $|\boldsymbol{k}|$ is near but not equal to zero. The direction of $\boldsymbol{k}$ is clearly somewhere between those of $\boldsymbol{k}_i$ and $-\boldsymbol{k}_s$. For photons, $|\text{wave vector}| = 2\pi \times$ refractive index $\times$ wavenumber, in the direction of propagation. For the blue–green region, wavenumber $\approx$20,000 cm^{-1}.

For acoustic modes, with a unique displacement vector associated with each cell, we see that the wave vector may be parallel to this displacement (longitudinal modes), or perpendicular to it (transverse modes), or in between. For infrared-active optical modes there is also a unique vector associated with each cell; this is the change of dipole moment during the vibration (more strictly the limiting direction at zero displacement). Such modes may also therefore be classified as transverse and longitudinal. In longitudinal modes all the dipoles over a half-wavelength reinforce one another and interact; for this reason longitudinal dipole-active modes always have a higher frequency than transverse versions of the same factor group fundamentals. The unperturbed or transverse modes are normally observed in infrared transmission measurements (though there are complications). For modes which are both infrared- and Raman-active (occurring only in non-centrosymmetric crystals) the Raman measurements may show either version, or a 'hybrid' according to the scattering geometry. This modification of the simple group-theoretical treatment must be considered when appropriate. By considering one direction for the wave vector in a crystal, in relation to degenerate dipole-active modes, it may be said that the degeneracy is split. This description is slightly misleading because there remains the 'degeneracy' obtainable from two different but equivalent experimental geometries with the same crystal, so that the wave vector is also 'turned'.

There is also the possibility of mixing of different symmetries of dipole-active modes under the influence of the dipole. For chemical purposes it is clearly always preferable to observe transverse modes, which (not *too* close to the zone centre) closely approximate to the frequencies determined by other than long-range electrostatic forces.

For a fuller discussion of this subject, we must re-examine what we mean by 'crystal symmetry'. In the sense that we have used it so far, this means simply the symmetry of the equilibrium positions of the nuclei. We have tacitly ignored possible disturbance of this symmetry by thermal population of vibrational levels, and by observations. We might also have to define the symmetry in other ways for other purposes, e.g. to take account of magnetic phenomena.[87]

For vibrational purposes it turns out that none of these considerations is important, except for the dipole moment vector associated with longitudinal or hybrid modes. This need be taken into account only when observing such modes; i.e. their thermal excitation is irrelevant (to overall *symmetry* properties). However, when such a mode is observed, it must be considered that this vector is present in the crystal when enumerating the crystal symmetry operations. Thus operations which do not send the vector into plus or minus itself may be considered lost in such circumstances. This is the reason for the mixing, and also the origin of the 'split degeneracy' description. It may be seen that the latter is no more than a restatement of the different frequency of longitudinal and transverse modes, in the special case of modes degenerate under the factor group.

Longitudinal and hybrid modes are best avoided by the chemist unless for special purposes. However, in case the reader wishes to pursue the subject, we fill in some of the common ground assumed by physicists when writing about the subject. The difficulty appears to be that physicists argue from very simple and highly symmetrical crystals, and regard anything else as a special case. To the practising chemist, of course, simple highly symmetrical systems are very special cases indeed!

The simplest possible case is a cubic material with one optical factor-group fundamental only, active in the infrared and Raman (there are a few non-centrosymmetric cubic structures). This vibration is considered in terms of positively and negatively charged sub-lattices vibrating against one another. It may be seen that in this case there is both a unique dipole moment and a unique displacement vector associated with each cell, and that they coincide; thus, transverse and longitudinal modes may be defined by use of either vector in such simple systems. Since the material is cubic, the sub-lattices may vibrate against one another in any direction, without change of frequency in the factor group approximation. The mode is therefore triply degenerate under the factor group. In fact, of course, the longitudinal version has a higher frequency than the transverse, so that *for one (arbitrary) direction of the wave vector* the degeneracy is split into one high-frequency longitudinal mode and a doubly degenerate transverse mode. For dipole-inactive crystal modes, it may be assumed that a Raman observation forces a triply degenerate mode into components defined by the Raman tensor, i.e. by the electric vector directions of exciting and observed light. Anyway, the same result appears whatever components are considered. However, in a Raman observation of a dipole-active triply degenerate mode, it must be considered that one component, the longitudinal, is defined *by the observation* with dipole and wave vector coincident. The other two (transverse) components may, however, still be taken at random in their plane. Thus, in general two bands appear in such a Raman spectrum; cf. cubic zinc sulphide, with Raman bands at $\Delta\nu =$ 276 and 351 cm^{-1}.[88]

For crystals other than cubic, dipole-active modes are regarded in the same way, as arising from relative displacement of a positive and a negative sub-lattice. Thus three different modes are regarded as simply different attributes of one pair of sublattices. This is misleading to the chemist, who is accustomed to regard different modes as different modes. It is also a difficult concept on which to generalize, since where more than one positive and one negative sublattice exist, the actual modes are mixtures of their relative displacements, different in different directions for non-cubic crystals. It will be seen that, with one pair of sublattices in non-cubic crystals, it is possible to formulate the problem as a three-fold degeneracy split in a particular orientation by the anisotropy of the force constants, with which (crystal symmetry) orientation the long-range electrostatic forces are, in general, in competition. It is possible to formulate the results in a usable way when either one effect or the other predominates. The case of many sublattices can be dealt with only by product rules (cf. isotropic substitution) involving all the dipole-active modes.

Where the three principal directions of the force-field ellipsoid are determined by crystal symmetry (orthorhombic or higher), it is possible to predict and, in principle, to use scattering geometries in which the wave vector lies along one of them, so that the long- and short-range forces are not in competition, and pure traverse or pure longitudinal modes are observed. In other cases this is not possible. Such cases have as yet been avoided even by physicists. Chemists are advised to temper their actions accordingly!

4.2.3.2 *Polaritons.* With forward scattering, the phonon wave vector is clearly much smaller than either incident or scattered photon wave vector, and, at some angle to the forward, the wave vector approaches the value for the photon resonant with the phonon (i.e. the infrared absorption frequency photon). The crystal disturbance corresponding to a transverse dipole-active mode of this wave vector is in fact part photon and part phonon, and has been called a 'polariton'.[89] For this reason the observed frequency drops as forward scattering is approached.[90]

4.2.3.3 *Linear polymers and line-group vibrations.* This problem is not different in principle from the crystal vibration problem. In one way it is much simpler because, for an 'isolated' polymer, the wave vector is restricted to the one direction in which a primitive translation occurs. For such a polymer in a crystal lattice we would normally be interested in the component of the wave vector in that direction. For real polymers we have to consider 'end effects'. Little information appears to be available on the length necessary for these to be unimportant. However, in practical terms it is not generally possible to obtain a material with a sharp molecular weight with more than about 10^2 units present. Such a material is, of course, necessary before specific length effects can be discussed. In an inhomogeneous material it is possible to consider only general departures from 'factor group' behaviour caused by statistically distributed end-effects.

The wave vector rule stated previously is still valid for infinite and cyclic polymers. Thus for N_p units in the polymer,

$$|\boldsymbol{k}| = 0,\ 2\pi/N_p,\ 4\pi/N_p,\ \ldots \tag{15}$$

However, this rule in its simplest form, where $\cos(rk)$ and $\sin(rk)$ standing waves are degenerate for a given k, applies strictly only to infinite and cyclic polymers. As we have discussed, this should not matter for large N_p, but for shorter chains we must expect slight modifications. In particular, the centre of the chain is, for centrosymmetric units, a symmetry element even of the chain considered as a molecule, and as such must be the position of a node or an antinode in all modes. To accommodate this, we allow the running parameter r to take the value zero at this point (if in a cell) or $+\frac{1}{2}$ and $-\frac{1}{2}$ on either side of it (if between cells). Clearly, the sine and cosine pairs are no longer strictly degenerate in these circumstances (belonging as they now do to different symmetry types of the chain considered as a molecule). The number of unit cells in the chain is not easily defined; considering the cyclic polymer, there is the question of the cell 'broken open' to make the linear version. Extra atoms also appear at the chain-end to satisfy the 'loose valencies'.

We cannot give here all the details of this process. We should note that it is not immediately very obvious that the detailed solutions to the secular equations for full chains of limited length are going over into the 'cyclic' rules for long chains. There is also the effect of thermal population of vibrational levels to consider. This whole subject is dealt with in greater detail by Zbinden for example (bibliography at end of this chapter). Here we summarize only the main features.

Applying our previous arguments, we may suppose that for $|\boldsymbol{k}|$ close to zero, with an 'infinite' chain, the factor group classification applies. For $|\boldsymbol{k}|$ removed from zero it is clear that a basic minimum requirement is that the overall point group of the polymer treated as a molecule must be usable.

For a 'very long' polymer held in a crystalline lattice, the overall selection rule $|\boldsymbol{k}| \approx 0$ applies. It is usual to treat high polymers in this way. Although they are often not obtainable as ideal crystals, except where they are groupings existing only in a crystal, they may be made to consist largely of crystalline regions. If the spectra of less- and more-crystalline samples are compared, it may also be possible to extrapolate to true crystallinity.

For low polymers, even when held in a crystalline lattice, the length is important. Considering a series of gradually increasing length, we may expect certain effects. For short species, many of the $|\boldsymbol{k}| \neq 0$ modes will give rise to prominent features of the molecular spectrum, classifiable according to the molecular point group. As the length increases, the $|\boldsymbol{k}| \neq 0$ modes reduce in intensity, finally dying out altogether. At the same time their spacing decreases, so that those few of one branch remaining in the spectrum finally coalesce. During this process, the factor group classification increasingly applies to the $|\boldsymbol{k}| = 0$ modes. At intermediate lengths, the factor group classification applies well to units near the centre of the chain, but near the ends mixing increasingly occurs, as allowed under the molecular point group. This mixing still occurs for very long chains, but it then applies to such a small proportion of the material that it may be neglected.

From these remarks, we see that it may be possible to extrapolate observed spectra both to perfect crystallinity and to infinite length. It may also be possible to discuss the spectra of intermediate-length and low polymers, in terms of a perturbed (end effects) and constrained (**k**-values) infinite polymer in which the $|\mathbf{k}| \approx 0$ selection rule is breaking down or has disappeared.

We have seen that the rule governing possible **k**-values of vibrations (whether or not spectroscopically observed) requires re-formulation in the presence of end effects. For instance, with the running parameter r re-sited to be zero at the chain centre, a better rule corresponding to the 'free end' model (see Zbinden) allows $\mathbf{k}$ to take the values:

$$|\mathbf{k}| = 0,\ \pi/N_p,\ 2\pi/N_p,\ 3\pi/N_p,\ \ldots\ (N_p-1)\pi/N_p \qquad (16)$$

and constructs standing waves with sine and cosine functions alternately for the consecutive values of $\mathbf{k}$. N_p in this case is defined as the number of cells in a hypothetical cyclic chain, broken open to make the real linear chain. We note, however, that the chain ends do not move freely whatever the displacement, in a solid. The wave-vector rule for a chain with 'fixed' ends is:

$$|\mathbf{k}| = \pi/(N_p+1),\ 2\pi/(N_p+1),\ 3\pi/(N_p+1),\ \ldots\ N_p\pi/(N_p+1) \qquad (17)$$

These rules may only be applied where the motion in adjacent cells is similar enough to be able to define a space phase at all. For example, the normal-hydrocarbon 'accordion' modes studied by Schaufele (Chapter 7) are acoustic modes for which the spacing is determined by the possible **k**-values. In this case the dispersion is large (characteristic of acoustic modes) and discrete bands are apparently observed up to about C_{100}. It is almost impossible to obtain homogeneous materials for $C_{>100}$ in any case. For the 'longitudinal' modes of centrosymmetric chains, only the cosine functions are gerade and only the sine functions ungerade.

4.2.3.4 *Sub- and super-structures.* Crystal structures are sometimes encountered in which a primitive cell may be defined, which gives rise to a not quite perfect primitive translation operation. Often a larger cell (supercell) may be defined for which the operation is perfect. Conversely, if the structure is described with this larger cell, it is said to have a substructure. It is easy to see the theoretical effect of this situation on the crystal vibrations and selection rules. If we consider the subcell description, then for q cells laid end-to-end to form the supercell, slightly active modes may arise for

$$|\boldsymbol{k}| = 2\pi/q,\ 4\pi/q,\ 6\pi/q,\ \ldots \qquad \pi \qquad (q \text{ even}) \tag{18}$$

$$\text{or} \qquad \ldots \quad (q-1)\pi/q \quad (q \text{ odd}) \tag{19}$$

in addition to every $|\boldsymbol{k}| = 0$ mode, with the additional $\boldsymbol{k}$-values restricted to the direction of elongation of the cell. Acoustic modes with these $\boldsymbol{k}$-values become, in principle, optical modes.

The reader may verify that these $\boldsymbol{k}$-values, based on the subcell, are all $|\boldsymbol{k}| = 0$ waves for the supercell. To determine which of these modes may actually have slight activity, and whether in the Raman or the infrared, they must be classified according to the symmetry types of the superstructure factor group. This factor group also controls the mixing of modes. It should not be assumed that this mixing is small, since we are not at the zone centre for the substructure. However, the mixing of modes of different substructure $\boldsymbol{k}$-values, and of the substructure $|\boldsymbol{k}| = 0$ modes between themselves, will be small, since both of these are forbidden by a perfect substructure.

The reader who has followed this argument has also a valuable insight into the general nature of the symmetry properties of modes away from the zone centre, and of the way in which they might be classified. For example, Raman himself confines his discussion of second-order effects to a supercell of dimensions $2 \times 2 \times 2$ primitive cells. A full treatment must, however, go beyond this.

It is found, as usual, that for all the $\boldsymbol{k}$-values listed as acceptable, apart from $|\boldsymbol{k}| = 0$ and $|\boldsymbol{k}| = \pi$, there are two different choices possible under the superstructure factor group, according to whether the centre of the supercell is chosen as a node or an antinode of the substructure standing wave. Both choices are acceptable, and this accounts for the $3nq$ superstructure factor group fundamentals.

It might be objected that this is making two crystal modes out of one. However, we are dealing with an effectively infinite crystal, in which the bands in the observed first-order spectra are the envelope of a number of crystal modes, close to the zone centre rather than at it, and the permitted values of the wave vector are so close together that it is effectively a continuous function. Under these conditions there are clearly more than enough crystal modes. The formal resolution of the paradox lies in the slightly less narrow spacing of the permitted wave vector values in the superstructure description, which has only N/q unit cells in a given piece of crystal.

Very little vibrational spectroscopy appears to have been undertaken for structures of this nature. However, it seems that vibrational methods are very much less sensitive to this type of distortion than are X-ray methods. This is partly because the appearance of even very weak X-ray reflections with formally non-integral indices is an immediate and unambiguous indication of a superstructure. The extra bands appearing in the vibrational spectra might, however, arise from other causes—impurity or second-order processes. It appears also that such bands would often be masked by second-order effects. In principle they might be distinguished by their sharpness, and possibly by low-temperature studies, although the superstructure may disappear at low temperatures in some cases.

One most important paper by Bauman and Porto has appeared.[91] In this they were able to show, by means of single-crystal Raman and infrared studies, that the tysonite structure, as exhibited by certain rare-earth trifluorides, exhibits gross features explicable under a bimolecular D_{6h}^4 unit cell. However, they showed that the true six-molecule unit cell, already detected by X-ray diffraction, had not the predicted D_{6h}^3 symmetry but rather D_{3d} point symmetry, uniquely indicating, in this case, D_{3d}^4 space group. This confirmed a result obtained from optical absorption spectroscopy on U-centre materials. The materials had to be cooled to $\sim$70°K to obtain the Raman results.

4.2.3.5 *Disorder.* In the detection and characterization of the more general types of disorder, the balance probably turns from diffraction methods in favour of vibrational spectroscopy. Statistical disorder without any symmetry properties may be considered as of several types. Deviation of atoms from an absolutely regular position in the structure is found only as an abnormally high thermal amplitude in X-ray determinations. In the vibrational spectra, we would in principle expect the factor group selection rules to break down, particularly for vibrations involving much motion of an 'offending' atom. Very large, perfect crystals are necessary for measurements of the refinement that would be required to reveal this. In favourable cases it might be possible to achieve this.

Another type of disorder which might arise, the distribution of two atoms of similar mass randomly into two positions, would be expected to have a very great effect on the vibrational spectrum, where very different force constants arose involving these atoms. This argument does not, of course, apply to isotopic substitution. No determination of that kind is known to the authors. It should be noted that neutron diffraction is potentially a powerful tool for this type of determination. Several mixed crystal systems have been studied by Raman spectroscopy, but no structural information unobtainable by X-ray methods arises from these studies. It is, however, interesting to see what does happen to the vibrational spectra.

For a summary and list of theoretical treatments, see ref. 92, in which Chang *et al.* deal with the CaF_2–SrF_2 and SrF_2–BaF_2 systems. It appears that in such structures, where the variant atom lies at a formal symmetry centre and may, for a pure compound, not move in Raman-active modes, a single Raman band is observed at all concentrations, corresponding to each pure compound Raman band. The frequency varies smoothly with concentration, and the half-width is at a maximum for a 1 : 1 mixture. However, where the variant atom does move in Raman-active modes of the pure compound, as in ZnS–ZnSe (ref. 93) and CdS–CdSe (ref. 94), it may be considered that, at small impurity concentration, a weak 'local mode' appears corresponding to the foreign atom. In the middle or mixed crystal region of concentration, the behaviour is not incompatible with some mixing of separate sulphur and selenium sublattice modes, complicated in this instance by the dipole activity of the modes studied. The complications arise only at finite impurity concentration, for the impurity sublattice, which is otherwise too dilute. Similarly, in the Si–Ge system[95] where no dipole is present for the pure component modes and no *resultant* dipole for a mixed crystal, the vibrational results do not really distinguish between true mixed crystals (which are undoubtedly present) and a collection of pure zones. The only effect noticed is a slight shift in the one Raman-active fundamental arising from each pure element.

Any system in which the primitive translation operation is made imperfect may, of course, be expected to exhibit first-order Raman spectra for modes not at the formal zone centre. Such bands are likely to be broad because of dispersion, and in general they are more easily observed for structures lacking a first-order Raman spectrum in the perfect version (so that the effects are not obscured). For example, Raman scattering observed in alkali halides containing F-centres[96] appears to originate from non zone centre phonons. Similar results are observed for defect BaF_2 and SrF_2,[97] and in all these cases it is probable that the colour appearing in the crystals enhances the Raman scattering efficiency. In cases where the colour centres are due to substitution of one ion by another, e.g. U-centres which arise from replacement of some X^- by H^-, local modes may also appear as in the mixed crystal systems in the 'slight impurity' concentration region.

4.2.3.6 *Structure determination.* Although X-ray methods are normally unambiguous, there are a few situations difficult to examine by X-ray diffraction, which may be accessible to vibrational studies. Probably the most important of these occurs when two atoms of nearly equal mass appear in a structure in such a way that the two possible arrangements are equally acceptable chemically. Here the probable difference in force constant favours the vibrational approach. In some cases single-crystal studies may not even be necessary. For example, the Raman spectrum of polycrystalline WOF_4 clearly demonstrates the presence of a terminal oxygen atom,[98] so the suggested oxygen-bridged structure cannot be correct. A related problem, again accessible to neutron but not to X-ray diffraction, is the position of light atoms in a structure. In favourable cases this can be determined unambiguously, but vibrational methods are a poor alternative to neutron diffraction, partly because the vibrational methods are effectively restricted to symmetry arguments. These remarks apply not only to 'absolutely' light atoms but to any situation where there is a large disparity in mass, e.g. in heavy-metal oxides. A classic example is the determination by Mathieu of the configuration of the water molecules in $LiClO_4,3H_2O$.[99]

In principle, single-crystal vibrational spectroscopy can be used to determine crystal class (i.e. point symmetry) by empirical determination of the selection rules which appear to be operating. The tysonite fluoride results of Bauman and Porto[91] illustrate this point.

4.2.3.7 *Chemical bonding and the site group.* It is a matter of experience that in crystals, as in fluids, there are 'strong' or 'chemical' bonds holding some atoms together into groupings, whilst between these groupings the forces (generaly known as van der Waals forces) are much weaker or even locally repulsive. In liquids the weaker bonds are not capable of maintaining long-term order. In crystals such order is (normally) maintained, but the distinction is just as real though less simply defined. Only occasional cases arise where accurately determined interatomic distances do not immediately reveal the nature of the interaction.

These remarks have to be partly modified for 'ionic' compounds and structures. We may, however, still distinguish 'covalent' bonding, which, for example, persists on dissolving such a compound, from the attraction of oppositely charged ions in a non-stereospecific way.

Many crystals consist simply of arrays of molecules or ions already well known in pure fluids or in solution. In others there are clear 'chemically bonded' groups which may extend infinitely in one or two directions, being effectively linear or planar polymers, although existing only in their crystal lattice. Finally, some crystals are simply infinite three-dimensional polymers.

The monomeric, less than three-dimensional polymeric species (and three-dimensional species which constitute only a part of the whole array) are to some extent vibrationally independent of one another, and of the rest of the crystal. This is simply because of the very much stronger forces acting within the groups than between them. As a result, many crystal modes may be classified as 'internal' (within such a group) or 'external' (where a whole group moves together but e.g. in the opposite direction to its neighbour). The external movements of individual ions subject to strong but non-stereospecific external forces are generally considered along with the external modes. Clearly the classification may break down badly in ionic crystals; e.g. there may be considerable coupling between external crystal modes and the internal modes of a polyatomic ion. Generally, however, the approximation is a good one. It has become customary to refer to external modes as 'lattice' modes, although strictly all crystal modes are governed by the periodicity and are therefore lattice modes. This terminology has arisen because crystal spectra often show clear sets of bands directly comparable to those obtained from species in fluids, with in addition a few well defined external and therefore 'lattice' modes.

It will be appreciated that the symmetry operations of a crystal fall into two categories with respect to a grouping of the type discussed. In the first category are those operations which cause only a rearrangement of the atoms within the grouping, or which transform the grouping into one 'primitively related' (i.e. obtainable from the first by one or a succession of primitive translations). This group of operations is called the 'site group'.[100] Clearly for molecules, and for one-, two-, and three-dimensional polymeric groupings, the site group is respectively a point group, a Line-group, a Plane-group, or a Space-group. Insofar as internal vibrations exist they are classified according to the site group or, where appropriate, its corresponding factor group.

The behaviour of 'subgroup vibrations' with respect to an overall group has already been outlined (section 4.2). Thus a local vibration, appearing in each of p equivalent but not primitively related groupings, gives rise to p vibrations classifiable under the overall group, corresponding to p allowed combinations. This multiplicity of crystal bands arising in theory from one local band is often not observed, particularly in molecular crystals. The splitting arising in this way is generally called 'correlation field splitting' or simply 'correlation splitting'. A crystal band observed in these circumstances is called a correlation multiplet (doublet,

triplet, etc.) if the splitting is observed or is to be referred to. We note that apparent absence of mutual exclusion under the site group is not carried over into a centrosymmetric crystal, Raman and infrared activity residing in different components of a correlation split multiplet.

The prediction, using the site group, of the number and activity of internal modes is always reliable. No information is obtained from site group analysis about mixing with external modes, or with internal modes of other groupings, or about the 'correlation splitting'.

The rules governing the integration of the site group analysis into the overall factor group are exactly the same as for any set of sites in any overall group. In this case, however, the overall group symmetry species arising from one site species are more interesting than a detailed knowledge of the allowed combinations. These species are determined by the correlation theorem, which we consider later.

4.2.3.8 *Factor group analysis.* Factor group analysis is not, in principle, different from point group analysis. It may nevertheless be of use to discuss some of the pitfalls which arise, and to summarize the process to save recourse to one of the many volumes dealing with point group analysis.

The meaning and determination of the factor group have already been discussed (section 4.2.1.2). The important point to remember is that the factor group is derived from the space group by setting the primitive translation operation everywhere equivalent to the identity operation. It follows that an atom transformed by a factor group operation, into another related to the first by one or a succession of primitive translations, is regarded as invariant under that factor group operation. There are often several apparently different situations in which a single factor group symmetry element appears in the structure. If the above rule is followed, the same result should be obtained for any of them, in relation to a particular set of atoms in the primitive cell.

(a) *Identification of symmetry elements.* The first step in the factor group analysis of a given structure is to find what are the symmetry operations of the factor group. The isomorphous point group gives the kind and number of operations, but a crystallographic text must be consulted to determine the substitution of non-translational by translational elements. *International Tables for X-Ray Crystallography* is an invaluable reference for this work. In Volume I the 'full' form of the space group, listing all important symmetry elements, is given for each of the 230 possible space groups at the top centre of the page describing the properties of the space group. Conventions define a specific order in which symmetry elements parallel (axes) or perpendicular (planes) to conventional crystal axes are given. These are given in Table 4.2 for the thirty-two crystal classes (point symmetries). The axes given for the monoclinic, trigonal, and hexagonal classes are non-orthogonal crystal axes, and are *not* those used in the character tables.

Having determined what elements are present, they must be identified in the structure. This is not entirely straightforward, since there is no necessity for the elements to pass through a common point. *International Tables* are again an invaluable aid, since for all except cubic classes they show the position of the elements in the conventional unit cell. For centred cells it is also necessary to identify a set of primitive translations. It will not be found necessary to draw in the contents of an entire primitive cell, provided that the primitive relationships of the centred cell are determined in this way (see Table 4.1).

At this stage it is instructive to determine the subsets of the factor group which are the site groups (or site factor groups) of any polyatomic groupings present. For centred cells, in some cases it may not be possible to choose the elements of a site factor group, which is a subset of the overall factor group, in such a way as not to involve transformation of one grouping into a primitively related one. In that case it is more convenient to use the equivalent 'self contained' site factor group obtainable by replacing some symmetry operations by translational equivalents; see the case of $(NH_4)_2CuCl_4$.[101]

We now proceed to describe simple factor group analysis, without referring immediately to site groups. We use the symbolism of Mitra[102]. The first step is to determine ω_R, the number of atoms in the primitive cell invariant under symmetry operation R. The writer has found it convenient to redefine ω_R as the number of sets of primitively related atoms not transformed into another set by the symmetry operation R. With this definition, centred cells and split atoms should lose their terrors. Any cell or any portion of a structure may be considered, so long as it contains as much as, or more than, the material appearing in the primitive cell. In this portion the three primitive translations (or more than three if desired) are identified, and after this all sets of primitively related atoms are easily picked out. The number of sets should, of course, be n, the number of atoms in the primitive cell.

The procedure now follows that for point group analysis, except for the factoring of translations and rotations. The character of the operation R, in the representation of the factor group defined by the $3n$ cartesian

Table 4.2 Conventional Order of Presentation of Symmetry Elements for the Thirty-two Crystal Classes

Point symmetry		Order of presentation of crystal axes		
C_1	triclinic	No symmetry		
C_i	,,	Centre of symmetry only		
C_2	monoclinic	x	y	z
C_s	,,	x	y	z
C_{2h}	,,	x	y	z
D_2	orthorhombic	x	y	z
C_{2v}	,,	x	y	z
D_{2h}	,,	x	y	z
C_4	tetragonal*	z		
S_4	,,	z		
C_{4h}	,,	z		
D_4	,,	z		
C_{4v}	,,	z	x	$(x + y)$ (diagonal)
D_{2d}	,,	z	x	$(x + y)$
D_{4h}	,,	z	x	$(x + y)$
C_3‡	trigonal†	z		
C_{3i}‡$\equiv S_6$	,,	z		
D_3§	,,	z	x	$(x + 2y)$
C_{3v}§	,,	z	x	$(x + 2y)$
D_{3d}§	,,	z	x	$(x + 2y)$
C_6	hexagonal†	z		
C_{3h}	hexagonal	z		
C_{6h}	,,	z		
D_6	,,	z	x	$(x + 2y)$
C_{6v}	,,	z	x	$(x + 2y)$
D_{3h}	,,	z	x	$(x + 2y)$
D_{6h}	,,	z	x	$(x + 2y)$
T	cubic¶	z	$(x + y + z)$	
T_h	,,	z	$(x + y + z)$	
O	,,	z	$(x + y + z)$	$(x + y)$
T_d	,,	z	$(x + y + z)$	$(x + y)$
O_h	,,	z	$(x + y + z)$	$(x + y)$

*For tetragonal classes $x \equiv y$, and $(x + y) \equiv (x - y)$. Point group and crystal axes are the same.

†For trigonal and hexagonal classes, $x \equiv y \equiv -(x + y)$ (all at 120°) and $(x + 2y) \equiv (x - y) \equiv -(2x + y)$ ('interleaving' the first set).

‡Hexagonal axes.

§Hexagonal axes. For these point symmetries, only one triplet of equivalent directions perpendicular to z may be symmetry elements. Convention puts these directions as x, y, $-(x + y)$ for some space groups, but $(x + 2y)$, $(x - y)$, $-(2x + y)$ for others. This choice is made to result in the smaller of two possible unit cells, and also distinguishes certain pairs of space groups having the same symmetry elements differently distributed

coordinates [and therefore the $(3n-3)$ vibrational coordinates plus the three translations] is:

$$\chi_j'(n_i) = \omega_R(\pm 1 + 2\cos\Phi_R) \tag{20}$$

where Φ_R is the rotation angle for the operation R. The sign is positive for a proper rotation and negative for an improper rotation; all symmetry operations are classifiable under these categories if 'no rotation' ($\Phi_R = 0$) is included. For convenience we tabulate the function $(\pm 1 + 2\cos\Phi_R)$ in Table 4.3 for crystallographically acceptable operations. This table also provides a quick reference for comparison of conventional point

Table 4.3 Tabulation of the Function
$\pm 1 + 2\cos\Phi_R$
for the Crystallographic Symmetry Elements

		Proper Rotations	
Φ_R	*R* point symbol	*R* space symbol	$1 + 2\cos\Phi_R$
0	E	1	3
π	C_2	(2), (2_r)	−1
$2\pi/3$	C_3^1	$(3)^1$, $(3_r)^1$	0
$4\pi/3$	$C_3^2 \equiv C_3^{-1}$	$(3)^{-1}$, $(3_r)^{-1}$	0
$\pi/2$	C_4^1	$(4)^1$, $(4_r)^1$	1
$3\pi/2$	$C_4^3 \equiv C_4^{-1}$	$(4)^{-1}$, $(4_r)^{-1}$	1
$\pi/3$	C_6^1	$(6)^1$, $(6_r)^1$	2
$5\pi/3$	$C_6^5 \equiv C_6^{-1}$	$(6)^{-1}$, $(6_r)^{-1}$	2
		Improper Rotations	
Φ_R	*R* point symbol	*R* space symbol	$-1 + 2\cos\Phi_R$
0	σ	*m,a,b,c,d,n*	1
π	$i \equiv S_2$	$(\bar{1})$	−3
$2\pi/3$	S_3^1	$(\bar{6})^{-1}$	−2
$4\pi/3$	$S_3^5 \equiv S_3^{-1}$	$(\bar{6})^1$	−2
$\pi/2$	S_4^1	$(\bar{4})^{-1}$	−1
$3\pi/2$	$S_4^3 \equiv S_4^{-1}$	$(\bar{4})^1$	−1
$\pi/3$	S_6^1	$(\bar{3})^{-1}$	0
$5\pi/3$	$S_6^5 \equiv S_6^{-1}$	$(\bar{3})^1$	0

in space. For those space groups, prefixed *R*, which are conventionally described with rhombohedral axes, the former convention is always chosen for the alternative hexagonal axes, and no third symmetry element is therefore listed. Where three elements are listed for a trigonal space group, one is always the element (1) or 'no symmetry'.

¶For cubic classes, $x \equiv y \equiv z$; $x \pm y \pm z$ are also all equivalent, as are $x \pm y$, $x \pm z$, and $y \pm z$. Point group and crystal axes are the same.

group and space group symbolism. The number of degrees of freedom classifiable according to each of the irreducible representations of the factor group (or point group) is given by the reduction formula:

$$N_k = \frac{1}{N}\sum_j h_j \chi_k{}^*(R_j)\chi_j{}'(n_i) \qquad (21)$$

where N is the order of the group, or the total number of symmetry operations therein (individual members of classes counted separately), h_j is the number in the j th class, and $\chi_k{}^*(R_j)$ is the (complex conjugate of the) character of the j th class of operations in the irreducible representation k.

The characters $\chi_k(R_j)$ are given in Appendix 2 for a number of important point groups, including all crystal point symmetries. For the crystal classes C_4, S_4, C_{4h}, C_3, S_6. C_6, C_{3h}, C_{6h}, T, and T_h it is essential to take account of the reducibility of the two-dimensional representations using complex numbers. Complex characters are given in the appendix for the crystal and other point groups exhibiting 'separable degeneracy'.

The numbers N_k so obtained for factor group analysis are the numbers of optical plus acoustic branches originating at the zone centre, classified according to the factor group and therefore the zone centre symmetry. The acoustic branches result from the three primitive translations of the unit cell; these may therefore be removed by subtracting from N_k the number of translations classifiable under symmetry type k (see character tables). The remainder represent the optical or factor group fundamentals observable spectroscopically, according to the point group selection rules. In Table 4.4 the factor group analysis for anatase is given;[103] the centred unit cell for this structure is shown in Figure 4.3.

Mitra[102] gives a method by which much of the site analysis is carried out as part of the factor group analysis. The interested reader may refer to the literature, but we give here a more descriptive treatment which has been found useful and quick, and equally applicable to site factor groups.

It is possible to start with a simple analysis of the grouping, considered under its site group (or the factor group thereof). Often, however, the grouping has, or may be supposed to have, a rather higher symmetry in the free state. In such cases it is generally more useful to analyze it under this higher symmetry (retaining rotation and translation), and to correlate this with the site symmetry in the usual way; see Appendix 2 and e.g. Wilson, Decius, and Cross[2] (Appendix X-8: "Correlation tables for the species of a group and its subgroups"; reproduced in Appendix 2).

This information is also contained in the character tables. Cases

Table 4.4 Factor Group Analysis for Anatase, TiO_2

D_{4h}	E	$2C_4$	$2S_4$	C_4^2	$2C_2$	$2C_2'$	σ_h	$2\sigma_v$	$2\sigma_d$	i
D_{4h}^{19}	E	$2(4_1)$	$2S_4$	4_1^2	$2(2/m)$	$2(2/d)$	a	$2m$	$2d$	i
$\chi_j'(T)$	3	1	−1	−1	−1	−1	1	1	1	−3
ω_R	6	0	2	6	0	2	0	6	0	0
$\chi_j'(n_i)$	18	0	−2	−6	0	−2	0	6	0	0
$h_j\chi_j'(n_i)$	18	0	−4	−6	0	−4	0	12	0	0

$N_k(a_{1g}) = 1/16\,(18 - 4 - 6 - 4 + 12) = 1$ $\quad N_k(a_{1u}) = 1/16\,(18 + 4 - 6 - 4 - 12) = 0$

$N_k(b_{1g}) = 1/16\,(18 + 4 - 6 + 4 + 12) = 2$ $\quad N_k(a_{2u}) = 1/16\,(18 + 4 - 6 + 4 + 12) = 2$

$N_k(b_{2g}) = 1/16\,(18 + 4 - 6 - 4 - 12) = 0$ $\quad N_k(b_{1u}) = 1/16\,(18 - 4 - 6 + 4 - 12) = 0$

$N_k(e_g) = 1/16\,(36 + 12) = 3$ $\quad N_k(b_{2u}) = 1/16\,(18 - 4 - 6 - 4 + 12) = 1$

$N_k(a_{2g}) = 1/16\,(18 - 4 - 6 + 4 - 12) = 0$ $\quad N_k(e_u) = 1/16\,(36 + 12) = 3$

Raman: $a_{1g} + 2b_{1g} + 3e_g$; inactive b_{2u}; acoustic $a_{2u} + e_u$

Infrared: $a_{2u} + 2e_u$

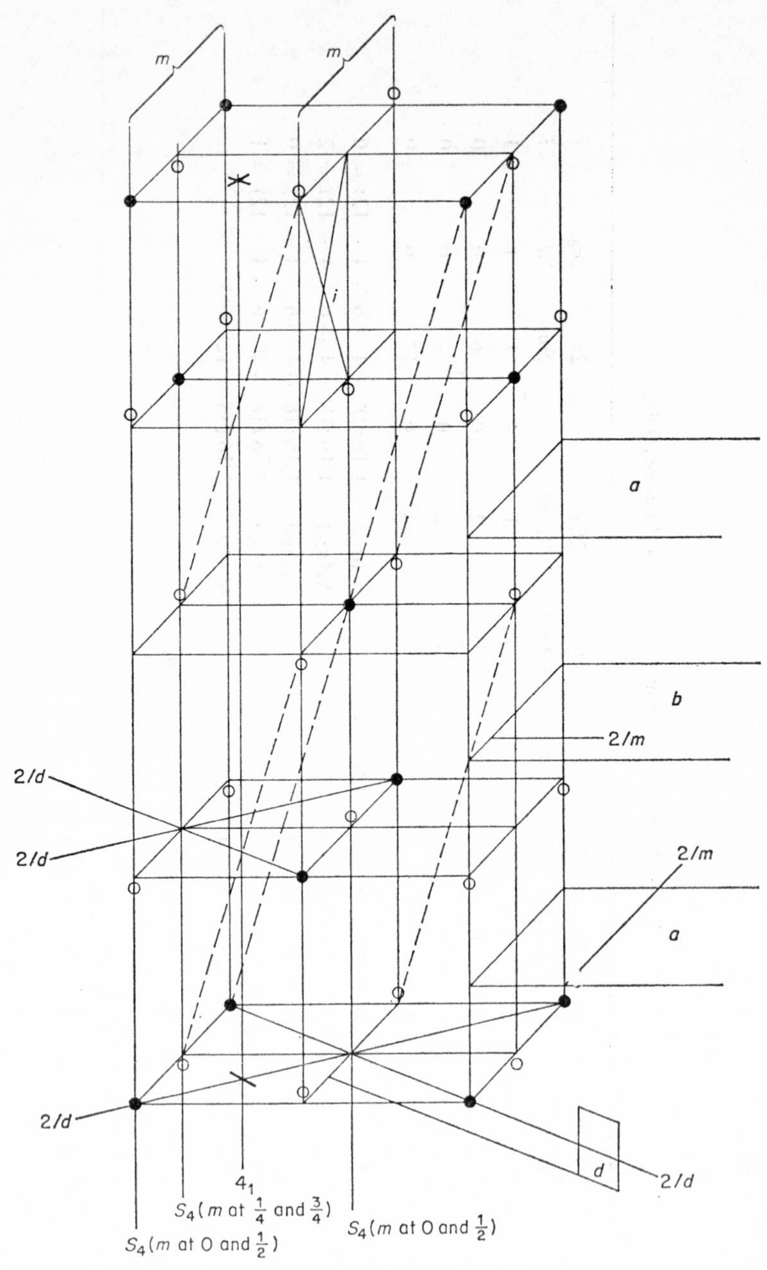
m
m
i
a
b
2/m
2/d
2/d
2/m
a
2/d
d
2/d
4_1
S_4(m at $\frac{1}{4}$ and $\frac{3}{4}$)
S_4(m at 0 and $\frac{1}{2}$)
S_4(m at 0 and $\frac{1}{2}$)

frequently arise where the results of this correlation depend formally on the choice of axes in the two groups, and it is strongly recommended for crystal work that the correlation be determined from first principles. This often saves confusion, and grouping axes coinciding with crystal axes may retain the name of the crystal axis. In the character table for the higher symmetry, blank out those columns corresponding to lost symmetry elements. For columns corresponding to classes of which the members are no longer in the same class of the subgroup, duplicate appropriately. All those species of the higher group now having the same set of characters become the same species of the subgroup. For one-dimensional and unsplit multidimensional species, the new species is simply determined by reference to the subgroup character table. For this purpose imaginary characters may be summed to give a two-dimensional species. Multi-dimensional species not classifiable under the subgroup in this simple way are split (reduced) into species of fewer dimensions in the subgroup. Summation of the appropriate rows of the subgroup reproduces the characters of these. In this way the subgroup species can quickly be found by intelligent trial and error. Alternatively the reduction formula may be applied.

Certain difficulties arise when the free grouping is linear and belongs to an infinite point group. Replacement of a molecular infinite-fold axis by a crystallographically acceptable 3-, 4-, or 6-fold axis is vibrationally irrelevant, and in such cases the grouping analysis should be carried out in the first place under the appropriate finite group. The molecular rotation about the axis is of course discarded. Hg_2Cl_2 is such a case.[103]

In other cases a similar technique may be used, carrying out the preliminary point group analysis under the finite subgroup of the infinite group, in which the infinite-fold axis is replaced by, say, a fourfold axis. It is, however, not so simple to deal with molecular rotation about this axis, when it disappears altogether in the site, and the molecule or ion is greater than diatomic. If it becomes a twofold axis in the site, or the molecule is diatomic, clearly the linearity must be retained and the molecular rotation may be discarded as before. When the axis disappears, *exact* linearity cannot be retained, and this corresponds to a 'free molecule' situation in which vibration and rotation may not be separated in any approximation. The correlation applicable to a large deviation from linearity is simply that one component of a degenerate deformation

Figure 4.3 Body-centred unit cell and symmetry elements for anatase, TiO_2. $D_{4h}^{19} = I\,4_1/amd = I\,4_1/a\ 2/m\ 2/d$. ● = Ti, ○ = O. Broken lines show primitive relationships

vibration becomes the 'missing' rotation. What actual spectra are likely to appear is another matter. If movement in the crystal became that of a hindered rotator, very complex effects could appear. This situation would be comparable to the hindered internal rotation of molecules. No work relevant to this situation appears to have been performed, but some hindered spherical and symmetric top rotators have been examined.

(b) *Correlation of site and factor group species.* Having classified the vibrations, rotations, and translations of the grouping under the site group (or its factor group, called from now on simply the site group), we must correlate these results with the crystal factor group species. This is done by the correlation theorem in a process which is effectively the exact reverse of the previous correlation from 'free' into site symmetry. The correlation tables may be used in reverse, or the character tables may be used as before. For a formal statement of the correlation theorem, see e.g. Wilson, Decius, and Cross.[2] In this case the factor group has the higher symmetry, and the site group is the subgroup. Each mode of the grouping becomes p factor group fundamentals of the crystal, where p is the number of groupings (or grouping primitive cells) per crystal primitive cell.

It is strongly recommended that the following section be read with a set of correlation tables to hand. Illustrations of the points mentioned may then be picked out by the reader. In uncomplicated cases it will be found that p factor group species correspond to one site group species; these are the symmetries of the p factor group modes, appearing as the p different allowed combinations (6) and (7) of one fundamental of the grouping.

Where exactly p such factor group species are not found, the situation is a little more complicated, because the allowed combinations of internal modes appearing in crystal modes are not entirely symmetry determined. If *every* appearance of a site species (of any dimensions) in a factor group species is counted D_f times, where D_f is the number of dimensions of the factor group species, it will always be found that D_s dimensional site species have been counted pD_s times.* This is true whether or not the simpler criterion (p appearances without reference to dimensions) is satisfied; the methods to be described are generally valid, and may be applied to that special case.

We consider first a simplified description of the correlation, and then point out the nature of the ambiguity arising where symmetry alone does not uniquely determine the result. For one-dimensional site species, the p allowed combinations are simply assigned as one to every one-dimensional factor group species in which the site species appears, and degenerate

*Note, however, that separably degenerate species (indicated by asterisks in the correlation tables in Appendix 2) must be treated as two separate one-dimensional species.

pairs or triplets for every appearance in a two- or three-dimensional factor group species. It will be recalled that cosine and sine functions for the same (space) phase difference are always a degenerate pair, and these may constitute both parts of a two-dimensional species, or two parts of a three-dimensional one. It should be pointed out that the division into two or three components is with a quite arbitrary orientation. All sets of orthogonal linear combinations of these components are equally acceptable orientations.

Where the site group species is also multidimensional, we are concerned with the pD_s combinations obtainable from the D_s components of every site mode. Every appearance of such a multidimensional site species in a (necessarily) multidimensional factor group species ($D_f \geqslant D_s$) corresponds to the appearance of the appropriate degenerate pair or triplet amongst the pD_s combinations. These results are a possible but idealized description of the (zone centre) crystal modes even after allowing for the approximate nature of the division into internal and external modes. This ideal may be likened to the description of molecular modes by single symmetry coordinates, which is, however, generally less valid. Where crystal modes of the same symmetry arise amongst the allowed combinations of site modes of different symmetry, the actual crystal modes may in principle be a mixture of those arrived at by the methods discussed. This may seem to invalidate the site analysis, but that is not so. We recall our previous remarks concerning atom sites (section 4.2). The crystal modes involved are always degenerate, and it is always possible to take components in such a way that, in any one component, the site modes of different symmetry appear in different sites of the set. However, different component orientations may be needed for different sets. It is intuitively apparent that such crystal modes will only be appreciably mixed where the site modes are very close in frequency.

Another ambiguity arises when a site species appears in the correlation table twice (or three times) in one factor group species. In this case two (or three) pairs or triplets of allowed combinations of one site mode become different crystal modes of the same symmetry. Any allowed mixture of allowed combinations may be expected in this case. It seems probable that the correlation splitting would also be unusually pronounced, being enhanced by an effective resonance mechanism.

We should also note some other ways in which the site analysis might, in principle, be misleading. For even the simplest relationship between site and factor symmetry, there may be more than one mode of the grouping classifiable under the same site group species. The mixing possible in this case is, of course, merely a restatement of the general effect of crystal forces on the form of the normal modes of the grouping. If, however, there is more than one set of groupings (whether or not arising from chemically distinct 'free' entities) the internal modes of these may mix in crystal modes of the same symmetry. Once again we expect this to be serious only for internal modes of similar frequency.

It will be recalled that the translations and rotations of the groupings were to be retained in the site group analysis. This is because they may be treated in exactly the same way as the true modes of the groupings, to arrive at crystal vibrations. The mixing phenomena will, of course, be much more marked for such external modes. By including the translational degrees of freedom of the remaining ungrouped atoms or single ions of the crystal in the same treatment, it will be seen that we arrive at a complete factor group analysis. The three translational degrees of freedom of the primitive cell (acoustic modes) must, of course, be subtracted from this result. It is in any case highly advisable to carry out a conventional factor group analysis to check for consistency.

It is not necessary for most purposes to know which of the p allowed phase differences corresponds to each crystal mode. For completeness we describe the (simple) method of obtaining this information, but most readers may ignore or only glance at this section. The first step is to find a symmetry operation R^1 by successive operation of which all groupings of the set are encompassed. In cases where such an element is not present, the set of groupings divides into subsets. One symmetry element relates the groupings of a subset, correlating into an intermediate site symmetry (which may if necessary contain translational elements). Another relates the subsets, now classified under their intermediate site group, into the factor group. Otherwise the treatment is identical, and we ignore this situation from now on.

The required correlation can now be determined by a method familiar in the generation of symmetry coordinates. The character of the generating operations R_r in the required factor group species is given for each allowed phase difference ϕ by:

$$\chi_{R_r} = \cos(r\phi) \tag{22}$$

for $\phi = 0$ and π, and by:

$$\chi_{R_r} = 2\cos(r\phi)^* \tag{23}$$

for $\phi \neq 0$ or π.

In applying equations (6), (7), (22), and (23), we must have a convention for the sign of a site vibration. We define two neighbouring site vibrations

---

*$e^{ir\phi}$ and $e^{-ir\phi}$ for separably degenerate species, using the corresponding imaginary components (4) and (5) (section 4.2).

to be in phase (and therefore of the same sign) when the symmetry operation relating the sites also brings the vibrations into exact coincidence. For 'handed' vibrations (vibrations of sites belonging to enantiomorphous site groups having no improper operations, or where the vibration belongs to a species having character $-D_s$ for all improper operations of the site group) an alternative convention may seem more logical. This defines the same sign for superposable vibrations, or the opposite sign for enantiomorphs. However, the linear combinations (6) and (7) must be taken with an alternating sign if this convention is used with an improper relating operation.

The integer r takes values 0 to $(p-1)$, and we define

$$R^0 \equiv E \tag{24}$$

It will be found that this information, *and* the limitation to the few factor group species determined by correlation tables or theorem, are together sufficient to determine the factor group species corresponding to a given value of ϕ and a given site species. Where greater than two-dimensional factor group species are involved, it will be necessary to sum characters appearing from two different values of ϕ. Where ambiguity arises, all indicated factor group species arise from the given value(s) of ϕ.

Examples. The process of site and factor group correlation is simple to carry out but less easy to describe clearly. We therefore illustrate the points involved with some examples. The species notation here and in Appendix 2 is that used by Wilson, Decius, and Cross.[2]

In the simplest possible cases the site and factor groups will be the same, so that a 1 : 1 correlation results. For example, in the K_2PtCl_4 structure, the $PtCl_4^{2-}$ ions have D_{4h} site symmetry, and the structure Space-group is D_{4h}^1. In this case D_{4h} is also the 'free ion' symmetry, and assignments were made very simply.[103] Similarly the K_2PdCl_6 structure has space group O_h^1 and O_h site symmetry $PdCl_6^{2-}$ ions.[103] We have already mentioned the Hg_2Cl_2 crystal, in which the D_{4h} site symmetry of the molecules is not vibrationally less restrictive than the $D_{\infty h}$ 'free' symmetry.

In general, where the 'free' symmetry of a grouping is either the same as its site symmetry or no less restrictive than it, assignments may be made directly under the factor group. Another case is the $Tl_2Cl_9^{3-}$ ion in $Cs_3Tl_2Cl_9$.[104]

In some other simple cases it may be possible to make assignments to internal modes using only the results of the correlation theorem, as in $CuCl_2,2H_2O$ and $(NH_4)_2CuCl_4$.[101] We consider the latter correlation in more detail later. In general, for more complex structures, containing several low-symmetry groupings in the primitive cell, it is necessary to use the oriented gas phase approximation as described in section 4.2.5. Meanwhile we give a few more difficult real and hypothetical examples of the application of the correlation theorem.

The first example is a real one, and concerns the correlation of distorted tetrahedral ions of C_s site symmetry, into an orthorhombic (D_{2h}) factor group, in crystals of e.g. Cs_2ZnCl_4 and Cs_2CuCl_4. The reader must consult the original paper[101] for details of the choice of axes, but it was required to correlate about the xz mirror plane of the factor group, this being the only operation applicable to the individual ions in the cell. Classification under the site group resulted in:

$$\Gamma_{\text{vib}} = 6a' + 3a''$$
$$\Gamma_{\text{rot}} = a' + 2a''$$
$$\Gamma_{\text{transl}} = 2a' + a''$$

Table 4.5 Correlation for D_{2h} and C_s Point Groups about σ_{xz}

D_{2h}	a_g	b_{1g}	b_{2g}	b_{3g}	a_u	b_{1u}	b_{2u}	b_{3u}
$C_s(xz)$	a'	a''	a'	a''	a''	a'	a''	a'

According to Table 4.5, every a' site mode becomes:

$$a' \rightarrow a_g + b_{2g} + b_{1u} + b_{3u}$$

in the crystal, and every a'' site mode becomes:

$$a'' \rightarrow b_{1g} + b_{3g} + a_u + b_{2u}$$

The correlation from T_d free symmetry into C_s site symmetry was also found using correlation tables in the more usual manner. There are four ions in the primitive cell, so 4 is the correct number of crystal modes to appear from each site mode.

In our second example, we consider a factor group, e.g. C_{4v}^3, with two-dimensional species, and a site group C_2 giving rise to resonance-enhanced correlation splitting. For this (hypothetical) example, the correlation is given in Table 4.6. We see that, for modes of species a under the site group, the associated crystal modes are:

$$a_1 + a_2 + b_1 + b_2$$

However, we see that all four linear combinations of one site mode of species b classify as various components of $2e$ species modes under the

Table 4.6 Correlation for C_{4v} and C_2 Point Groups

C_{4v}	a_1	a_2	b_1	b_2	e
C_2	a	a	a	a	$2b$

crystal symmetry. Since each e mode is doubly degenerate, the correct number of crystal vibrations results. The two crystal modes of the same symmetry may be expected to 'repel' one another, leading to enhanced correlation splitting of the b site modes.

For another example we return to the idealized $(NH_4)_2CuCl_4$ crystal.[101] For reasons probably of efficient packing, a sheet polymer of $(CuCl_4^{2-})_n$ in this structure (Figure 4.4), having intrinsic D_{4h}^5 Plane-group, is reduced

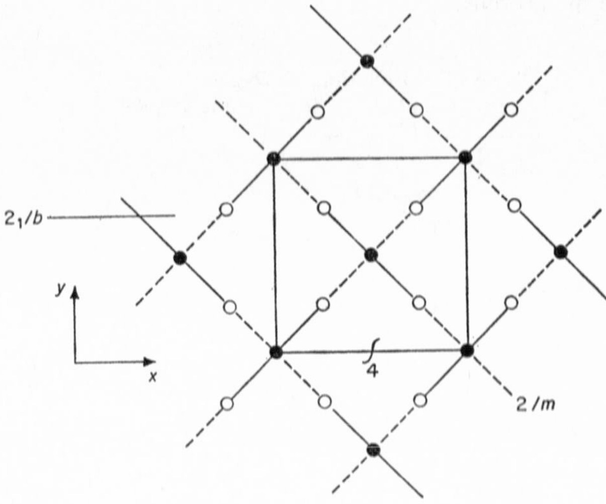

Figure 4.4 Single $(CuCl_4^{2-})_n$ sheet of the $(NH_4)_2CuCl_4$ structure showing axes and symmetry elements for the idealized isolated D_{4h}^5 polymer (i.e. $4/m\ 2_1/b\ 2/m$). (Additional terminal chlorine atoms lie above and below each copper. - - - - - - = 'weak' interaction; —— = 'strong' interaction; ● = Cu; ○ = Cl; ~shows position of one fourfold axis.) The sheet unit cell (outlined) contains the same amount of material as the structure primitive cell (apart from the cations) but the x-axis of the latter cell is not normal to the plane of the paper, so that successive sheets are staggered by half a cell

in the crystal to orthorhombic symmetry. The latter is Space group D_{2h}^{18}, but in order to describe the sheet $(CuCl_4{}^{2-})_n$ polymer with a self-contained Plane-group (not involving translation to a sheet primitively related in the crystal), it is analyzed under D_{2h}^{9} site Plane-group. However, we are concerned only with the correlation of $CuCl_4{}^{2-}$ ions, which, insofar as they exist, have D_{2h} site symmetry in the crystal, into the idealized D_{4h}^{5} factor group for the 'free' sheet. In this example many possibilities existed for the confusion of axes. Thus, using the D_{4h}^{5} standard axes, it turns out that z is along one pair of bonds of the square-planar discrete $CuCl_4{}^{2-}$ ion, and $(x+y)$ or $(x-y)$ is along the other pair, depending upon which of the symmetrically equivalent pair of ions in the unit cell is chosen. The twofold axis of the discrete $CuCl_4{}^{2-}$ ion in the plane is, however, along this second pair of bonds, which means that D_{2h} site and D_{4h}^{5} factor groups must be correlated around the diagonal C_2'' of D_{4h} point group rather than $C_2' = x$. Table 4.7 is the resulting correlation table, for gerade modes only.

Table 4.7 Correlation for D_{4h} and D_{2h} Point Groups around C_2'' of D_{4h}; Gerade Species Only

D_{4h}	a_{1g}	a_{2g}	b_{1g}	b_{2g}	e_g
D_{2h}	a_g	b_{1g}	b_{1g}	a_g	$b_{2g} + b_{3g}$

Two correlations are straightforward:

$$a_g \rightarrow a_{1g} + \quad {}_{2g}$$

$$b_{1g} \rightarrow a_{2g} + b_{1g}$$

However, for the rest:

$$b_{2g} \rightarrow e_g$$

$$b_{3g} \rightarrow e_g$$

so that, if a b_{2g} and a b_{3g} mode of the ion happened to be close together, one component of a resultant crystal mode might be a mixture of b_{2g} vibration from one ion and b_{3g} from the other. The other (degenerate) component would reverse this order.

In our next example (again hypothetical) we actually assign particular site phase differences to particular factor group species. The factor group concerned is $O_h{}^1$, identical with O_h point group, and the site symmetry C_{4v}. We note that the M–X bonds in regular octahedral MX_6 have C_{4v} site symmetry. We may therefore consider as our example six C_{4v} molecules distributed about a central point in octahedral fashion, in a primitive cell. The correlation of species is straightforward, and we merely give in Table 4.8 the result of detailed correlation of phase differences. The phases occur in S_6 space, this being the only one available which encompasses all six sites. The results according to the correlation theorem are:

$$a_1 \rightarrow a_{1g} + e_g + f_{1u}$$
$$a_2 \rightarrow f_{1g} + a_{1u} + e_u$$
$$b_1 \rightarrow a_{2g} + e_g + f_{2u}$$
$$b_2 \rightarrow f_{2g} + a_{2u} + e_u$$
$$e \rightarrow f_{1g} + f_{2g} + f_{1u} + f_{2u}$$

and Table 4.8 was constructed using this information and the formulae (22) and (23) for χ_{R_r}. Note that the six internal symmetry coordinates involving bond-stretching for MX_6 are obtainable from the results for a_1 site species. The twelve internal-angle coordinates are, however, only of C_{2v} local symmetry.

Where both site and factor group species are degenerate, some difficulty arises in orienting and assigning individual components. This ambiguity is indicated but not resolved in Table 4.8.

These methods apply also when the site group includes a primitive translation operation. In that case the two factor groups may be correlated as usual. For example, molybdenum trioxide contains sheet polymers with a site Plane-group of 'monoclinic' type, isomorphous with Space-group C_{2h}^2.[103] The site factor group C_{2h}^2 is correlated with the orthorhombic crystal factor group D_{2h}^{16}, using the $D_{2h}^{16}(P_{bnm})$ axes, for which the z-axis is the twofold axis of the sheet. Thus the conventional C_{2h} point group may be used for the latter. The resulting correlation is

$$a_g \rightarrow a_g + b_{1g}$$
$$b_g \rightarrow b_{2g} + b_{3g}$$
$$a_u \rightarrow a_u + b_{1u}$$
$$b_u \rightarrow b_{2u} + b_{3u}$$

In this case it was observed that correlation splitting of up to 50 cm^{-1} appears,[103] but nevertheless many Raman bands appear in clear-cut pairs $a_g + b_{1g}$ or $b_{2g} + b_{3g}$. Infrared dichroic measurements are also possible with these very thin sheet crystals.[105] It seems possible that the large

Table 4.8 Detailed Correlation for C_{4v} Site Group and O_h^1 factor group

	cos or sin $(r\phi)$						C_{4v} site species					
ϕ r =	0	1	2	3	4	5	a_1	a_2*	b_1	b_2	e_a	e_b
0, cos	1	1	1	1	1	1	a_{1g}	f_{1g}^a	a_{2g}	f_{2g}^a	$f_{2g}^{a'}$	$f_{1g}^{a'}$
$\pi/3$, cos	1	$\frac{1}{2}$	$-\frac{1}{2}$	-1	$-\frac{1}{2}$	$\frac{1}{2}$	f_{1u}^a	e_u^a	f_{2u}^a	$e_u^{a'}$	$\{ f_{1u}^{b'} + f_{1u}^{c'} +$	
$\pi/3$, sin	0	$\sqrt{3}/2$	$\sqrt{3}/2$	0	$-\sqrt{3}/2$	$-\sqrt{3}/2$	f_{1u}^b	e_u^b	f_{2u}^b	$e_u^{b'}$	$f_{2u}^{b'} + f_{2u}^{c'}$	
$2\pi/3$, cos	1	$-\frac{1}{2}$	$-\frac{1}{2}$	1	$-\frac{1}{2}$	$-\frac{1}{2}$	e_g^a	f_{1g}^b	$e_g^{a'}$	f_{2g}^b	$\{ f_{1g}^{b'} + f_{1g}^{c'} +$	
$2\pi/3$, sin	0	$\sqrt{3}/2$	$-\sqrt{3}/2$	0	$\sqrt{3}/2$	$-\sqrt{3}/2$	e_g^b	f_{1g}^c	$e_g^{b'}$	f_{2g}^c	$f_{2g}^{b'} + f_{2g}^{c'}$	
π, cos	1	-1	1	-1	1	-1	f_{1u}^c	a_{1u}	f_{2u}^c	a_{2u}	$f_{1u}^{a'}$	$f_{2u}^{a'}$

* 'Handed' site species.

correlation splitting may be characteristic of polymeric groupings in contact with each other in crystal sites.

(c) *Symmetry coordinates.* Since we are not concerned with vibrational analysis in this book we do not give an extensive discussion of the formation of symmetry coordinates. However, such coordinates are a useful way of discussing the vibrations of polymeric as of monomeric species. Symmetry coordinates for the zone centre vibrations of polymers may be generated under the factor group operations by conventional methods applicable to point groups. Often, however, the degree of redundancy for internal coordinates is so high that 'external' or cartesian symmetry coordinates are a more useful basis for discussion.[105] The relationship between cartesian and internal symmetry coordinates can then be deduced qualitatively in a sufficiently illuminating manner.

Such coordinates are often easily determined by inspection. We give a formal method also, since most of the treatment already exists in the preceding section on site group correlation. Every atom has its own site group, and its cartesians may be so aligned as to be classifiable thereunder. Symmetry coordinates are then generated as specified for the generation of crystal modes from site modes. The ambiguity concerning cartesians degenerate under the atom site group must be removed by consulting specialist treatments.

4.2.4 *Second-order Effects*

We have already indicated that second-order effects in solids tend to be diffuse and of indeterminate frequency. For this reason chemists at least are not likely to be much interested in them. This brief treatment indicates further the nature of the difficulties.

We define a second-order process in a solid as a two-phonon or a two-photon process. We have already considered the latter in the introduction, and we are here interested in two-phonon phenomena. We exclude the excitation of Raman scattering by light which itself originates as Raman scatter. However, it may be as well to remember this possibility when searching for very weak bands with powerful laser sources, in solids as in fluids. This phenomenon has also been treated in a unified way,[106] and it appears that observed intensity is proportional to the 4/3 power of crystal volume.

As remarked previously, simultaneous two-phonon Raman processes are not restricted to the zone centre, because the wave vector conservation rule need be satisfied only by the vector sum of the two phonon wave vectors. Likewise, some value of the wave vectors may often be found at which (weak) Raman activity appears for any pair of phonons. This activity is normally spread over a region of the zone, and so the bands are generally continua, depending on the dispersion of the frequencies through the appropriate regions of the zone. There may, however, be a sharp cut-off where the zone centre or zone boundary is encountered.

4.2.4.1 *Internal modes.* Where modes are largely confined to internal movements of monomeric species, we may expect slightly different behaviour. Overtone and combination bands for such species may be classified under the site group in the normal way, and crystal modes derived from them according to the rules given for fundamentals. A degree of broadening and of departure from the selection rules must be expected owing to coupling with external modes.

4.2.5 *The Oriented Gas Phase Approximation*

The basis of this very useful approximation is that, as some crystal vibrations may be derived directly from combinations of internal grouping vibrations, so may the crystal Raman tensor and infrared activity vector for such vibrations be derived from those applicable to the vibrations of the grouping. One would expect this approximation to be valid wherever the initial approximation, of division into internal and external modes, was also reasonable. However, the reader should beware of necessarily relating the degree of departure from the one to that from the other. Experimental evidence is limited, but indicates that the approximation works satisfactorily (if not exactly) for colourless molecular crystals. However, electronic transitions in the near-u.v. and visible must be treated with caution.

The oriented gas phase approximation was first introduced in the interpretation of the infrared dichroism of molecular crystals.[107,108] That treatment is very simply generalized. The exact physical significance of either the infrared or the Raman treatment is, however, more difficult to discuss. Whether, for instance, the observations or the predictions are truly those apertaining to an oriented gas phase (whatever that may be) might well be doubted. We therefore give the treatment but leave the reader to deduce the premises. The argument may be developed along the lines of the site group method of factor analysis. As in that case, there are two possible starting points, depending on whether there is a reasonable 'free' symmetry which may be assigned to the molecule, greater than the site symmetry. Where no such higher symmetry exists, the arguments may be stated in a very simple form. The *symmetry* limitations on the activity are exactly obeyed. The approximation may then be used to relate relative values of independent crystal tensor or vector components to those established e.g. by polarization measurements in fluids or *a priori* for the molecules. This possibility has not been much tested, and it must suffer from the ambiguity of measurements in fluids and *a priori* arguments. The mathematical treatment is contained in the general treatment to follow. We should also emphasize here that assignment to site group symmetry by single-crystal spectroscopy is unambiguous, not requiring any approximation. The only uncertainty lies in identifying crystal modes as internal modes.

Where molecules in a crystal site suffer formal but actually very small departure from an 'ideal' or 'free' higher symmetry, the 'gas phase approximation' is particularly useful. We find that the symmetry limitations on the molecular Raman tensor and infrared activity vector are in general approximately those of the higher symmetry, for the internal modes. We now relate molecular to crystal observations.

We must first clarify the symmetry limitations on the molecular tensor or vector. This is a matter of being very careful over choice of axes. In Appendix 2 we present point group character tables in which molecular x, y, and z axes are unambiguously defined in terms of principal symmetry elements. The classification of vector and tensor elements is made using those axes, or a stated rotation of them. The vectors are assigned to symmetry classes in the usual way. However, ambiguities of sign and normalization arise in the usual presentation of tensor components. We therefore adopt the method of Loudon,[86] presenting for each active species of every point group the Raman tensor in matrix form, using small letters to denote non-zero components. We remark later on the treatment of degenerate species. The symmetry limitations on β, the first hyper-polarizability, are also given, but in more compact form. (There is some doubt whether the oriented gas phase approximation will be usable for the hyper-Raman effect.) Where a symmetry element runs in common through idealized and site symmetry point groups and crystal factor group, it is clearly desirable to use it as an axis throughout, and to call it by the same name throughout. It is important to make the (simple) changes necessary to have this conformity. Often the names of axes merely need to be changed. Otherwise very simple transformations of axes are involved:

$$\boldsymbol{x}' = \boldsymbol{T}\boldsymbol{x} \tag{25}$$

in which case the new tensors are:

$$\boldsymbol{R}' = \boldsymbol{T}\boldsymbol{R}\boldsymbol{T}^{\mathrm{t}} \tag{26}$$

where T^t is the transpose of T, and the vectors are simply:

$$\mu' = T\mu. \tag{27}$$

In the following discussion it is assumed that such changes have been made.

Where molecular axes are not entirely determined by the site group symmetry elements, it is necessary to choose an orthogonal set as nearly related as possible to the (symmetry determined) free molecule axes. Similarly, some crystal structures (monoclinic and triclinic) are conventionally described in terms of non-orthogonal crystal axes, and it is necessary to choose an orthogonal set. We suggest some simple choices in section 4.3.5. Having chosen an orthogonal set of crystal axes, it is necessary to construct the transformation from orthogonal molecular to orthogonal crystal axes, for one molecule of the set. This is, of course, done by determining the nine direction cosines, although some relationships exist which lessen this calculation. The test of orthogonality in the molecular axes is the orthogonality of this transformation; it is possible at this stage to refine an approximately orthogonal set without needing to know the explicit angular relationships.

This transformation is applied to the Raman tensors and dipole moment vectors of the free molecule active species in turn. The result in each case is a vector or tensor with components in all or some of the places allowed through the series of p crystal modes associated with one molecular mode of the given site species. It is possible to show that the relative values of components, appearing in positions allowed by individual crystal symmetry species, are normally just the relative values for the individual crystal modes of those symmetries. This is true wherever such symmetry axes as are required to generate the p molecules from the typical one are used as crystal axes. One is thus saved the labour of applying a rotated transformation to p molecules in turn.

4.2.5.1 *Degenerate modes.* Some remarks on the oriented gas phase with respect to external modes appear in Appendix 2. Degenerate crystal modes fit into the scheme outlined without trouble, although we note the ambiguities discussed previously. The oriented gas phase approximation is thus defined by this treatment for all non-degenerate molecular or site modes.

Where degenerate molecular modes appear, we need to extend the discussion a little. For n-fold degeneracy, n separate but arithmetically equal tensors or vectors appear, with arbitrary orientation. Where the degeneracy is not split (or not observed to be split) the arbitrary orientation may be retained, each tensor or vector being treated as belonging to a separate species. Spectroscopically, an n-fold degenerate mode behaves as n modes of identical frequency, and predicted intensities can be worked out in this way by a summation after squaring.

Where a degeneracy in the free molecule is formally split in the site, it is desirable to be able to predict the tensor or vector applicable to each part. Where the degeneracy is split into site modes of the same symmetry, this is unfortunately impossible. The orientation of the separation is in such cases quite unpredictable. Where the site modes are of different symmetry, a quite simple rule may be stated. The split occurs in such a way that the resultant tensors or vectors are compatible with the site symmetry species indicated. The application of the rule may be more complicated. In principle orthogonal linear combinations (i.e. with scalar coefficients) of the given pair or triplet of tensors or vectors may be taken in such a way that the criterion is satisfied. As previously stated, the cartesian axes in use should coincide with important site symmetry elements, achieved by applying the same transformation of axes to each member of the given set. Some transformations are already indicated in Appendix 2. In general a degenerate pair will be oriented with respect to a particular one of a set of symmetry elements, and in Appendix 2 we note such relationships. The general requirement is to make the orientation with respect to an element retained in the site.

Comparatively few tests of the oriented gas phase approximation appear to have been made for Raman spectroscopy. Useful results have been obtained for $InCl_3$,$2NMe_3$ (ref. 109) and for Al_2Br_6 (ref. 103), and assignments made to the internal modes of naphthalene and anthracene.[110] The results for the latter two materials were, however, far from consistent with the predictions. Ga_2Cl_6 was also treated[103] by a method relying on the validity of the oriented gas phase approximation. The (triclinic) crystals (one molecule of Ga_2Cl_6 per unit cell) were simply examined in numerous different orientations, and groups of internal mode bands always showing the same behaviour were assigned to the same free molecule non-totally symmetric symmetry species. This was sufficient to assign the internal modes completely, after qualitative comparison with the relevant symmetry coordinates. Although it is in general inadvisable to treat coloured materials in this way, rhombic sulphur appears to behave satisfactorily,[111] using 6328 Å excitation. Success must presumably depend partly upon the relative importance of molecular interactions in the electronic wave functions. In examples with little interaction, any resonance enhancement might be expected to occur under the 'free molecule' symmetry, which would not matter. However, it is also necessary that the material should not be too strongly dichroic at the excitation frequency used; it would be very difficult to make absorption corrections for a solid.

Sodium dithionate $Na_2S_2O_6,2H_2O$ has been very carefully examined,[68] and the approximation found to be correct within experimental error. Furthermore, the solution depolarization ratios predicted from the crystal tensor were also within the experimental error of the observed values. In this case we should note that the crystal is a hydrate, so that the environment of the $S_2O_6^{2-}$ ion may not be very different in the crystal and in concentrated solution. The powder spectrum of this compound has been mentioned in Chapter 3.

There are some dangers in the use of the oriented gas phase approximation, even with electronically ideal materials. It is, of course, essential that the molecules or ions in their crystal site should not be subject to any gross disturbance of the form of their normal vibrations. Such disturbance may arise where two modes of the 'free' grouping, of different symmetry, lie very close together. If, in the crystal site, the symmetries become the same, extensive mixing may occur. This is almost certainly the explanation for the anomalous behaviour[101] of Cs_2ZnCl_4 in which $\nu_1(a_1)$ and one component of $\nu_3(f_2)$ of the tetrahedral $ZnCl_4^{2-}$ ion appear to mix in the crystal site, where both vibrations are totally symmetric under the site group.

4.2.5.2 *Bond polarizability changes.* It occurs not infrequently in using the oriented gas phase approximation that ambiguities remain because of independent non-zero Raman tensor elements for single modes under the 'free' symmetry. In such cases it may be possible to make qualitative use of the Wolkenstein hypothesis (Chapter 3). Thus, for nearly pure bond stretching vibrations, using the bond polarizability approximation, it is expected that the change of polarizability along the bond will generally be much greater than that across it. However, this may be less true for multiple bonds.

4.2.5.3 *Depolarization ratios.* Assuming the validity of the oriented gas phase approximation, it is in principle possible to obtain the mean value and anisotropy of the Raman tensor applicable to a 'molecular' vibration in a crystal, using single-crystal measurements, and without knowledge of the crystal structure. This allows immediate assignment to symmetric (mean value not zero) and non-totally symmetric (mean value zero) molecular modes. Intelligent comparison of the data with equations

(1)—(4) of Chapter 3 will often give at least a strong indication of such bands as have very small anisotropy (corresponding to strongly polarized fluid bands). Whilst a particular case may often be dealt with in some way derived from first principles, to fit it particularly, we note two possible exact general approaches:

(*a*) Where there is only one molecule in the primitive cell, and for certain other cases where correlation splitting cannot appear in the Raman, it is possible to make an unambiguous decision on the relative sign of diagonal crystal tensor components (section 4.4). This process may be complicated by birefringence (section 4.3.3). Equation (1) of Chapter 3 for the mean value may then be used directly; equation (2) for the anisotropy, not strictly necessary, provides a useful reference for gauging the comparative closeness to zero of the mean value, allowing for experimental error. It is essential to make use of such morphological, optical, and symmetry properties as may be known.

(*b*) The values of $\overline{R_{ii}{}^2}$ and $\overline{R_{ij}{}^2}$ [equations (3) and (4) of Chapter 3] can be found by averaging values obtained from a carefully 'randomized' selection of crystal orientations. Birefringence must again be considered (section 4.3.3), and it also precludes the use of other than cubic *powders* to obtain depolarization data (section 3.2).

4.3 Elementary Practical Considerations

4.3.1 *Relation of the Raman Tensor to Observed Intensities*

There are a number of papers in the literature which are concerned with this subject, e.g. ref. 103. The reader will find most topics from this point are treated more adequately in the literature than we shall have space for here. This is in contrast to the previous material, much of which is to be found either in over-concise mathematical form or in insufficiently general descriptive treatments.

We recall that the Raman tensor element R_{ij} controls the probability of scattering of light of electric vector direction *j*, where *i* is the direction of the electric dipole which appears to be radiating the Raman light. The intensity observed in solid angle increment $d\omega$ is therefore:

$$I \propto I_0 R_{ij}{}^2 \cos^2\theta \, d\omega \qquad (28)$$

where θ is the angle between $d\omega$ and the plane perpendicular to *i*. There is also ideally a dependence on the fourth power of the frequency of the scattered radiation, which does not at the moment concern us. I_0 is the incident intensity.

The propagation direction of the incident and scattered light is not relevant to this equation except for the definition of θ. It is not by any means certain that this is completely true in crystals; there are tentative suggestions that forward scattering may be slightly favoured even when incoherent.[56] The difficulties of measuring this are very great in crystalline materials. For normal practical considerations we may ignore the possibility. For dipole active modes we have seen, however, that it is most important to know both propagation directions, since these define the phonon wave vector direction with respect to the dipole direction. We are not further concerned with dipole active modes.

As stated before, the great advantage of single-crystal Raman spectroscopy derives from the ability to isolate the Raman tensor components. We recall from Chapter 3 that even a wide angle of collection introduces only a small and manageable intensity due to a second component. Using a notation developed by Porto *et al.*,[112] we write:[103]

$$I_{i(xy)j} = K(R_{yx}{}^2 + d_y R_{jx}{}^2)^* \tag{29}$$

The symbolism i(xy)j denotes excitation propagating along *i*, with electric vector along *x*; and collection along *j*, analyzed for vector along *y*. The divergence factor *d* allows for any system with wide-angle collection of at least twofold symmetry, where *j* is the twofold (monochromator) axis. For collection systems of lower symmetry, a third term in R_{xy} R_{xj} may be formally necessary.[103]

A more important consideration is the variation of the proportionality factor *K*, with the relation of *y* to the monochromator. All monochromators preferentially transmit light polarized in a particular direction. The factor involved, and the relation of the optimum direction to (e.g.) the grooves of the grating, vary with wavelength. A polarization scrambler may be used, but it is of course more efficient to keep the analyzer set at optimum, and to turn the crystal where necessary. One reputable manufacturer indicates wrongly the optimum (factory pre-set) analyzer direction.

Criticism has been directed at the Cary 81 'hemispherical' lens collection system whether used for 180° or 90° collection, because of the wide angle of collection, particularly for single-crystal work. In our experience this criticism is unjustified for chemical applications. A correction may easily be made for this error where required, and it is often much less than the errors introduced by crystal imperfections. An *f*/1 collection lens is

---

*Although strictly correct, this would often be written $R_{xy}{}^2$ etc.; so long as $R_{xy} = R_{yx}$ no confusion arises.

not anyway uncommon. Careful studies[68] of sodium dithionate $Na_2S_2O_6$, $2H_2O$, using a crystal about $3 \times 5 \times 10$ mm stuck to the hemispherical lens with Nujol, showed a divergence factor of $\sim$0·2. This compares with a theoretical factor of $\sim$0·04 (Chapter 3), and with the failure to detect the factor for liquids. The reason for this discrepancy is not known, and it has not been ascertained whether the factor always appears for large collection apertures or whether it is more a function of the hemispherical lens. A minor variation of d with refractive index, and therefore orientation, of an anisotropic sample is of course expected.

A general formula relating observed intensity to the Raman tensor, where the two sets of axes are not aligned with one another, is given by Loudon.[86] This formula is, however, subject to the limitations imposed by birefringence, and a more straightforward procedure is to apply a transformation to the Raman tensor. The formula would be of use in a theoretical evaluation of a finite-angle collection system. We also wish to state that there are distinct advantages in using 180° collection for difficult samples. It has been (erroneously) supposed in some quarters that 90° collection must be used for single-crystal work. There is no theoretical basis for this belief. The phonon wave vector for 180° measurements is twice that for 90°, but still so small as to be acceptably near the zone centre.

4.3.2 *Technique and Setting*

Techniques for single-crystal measurement vary widely according to the precision required and the nature of the sample. Undoubtedly the most difficult part of the operation for most samples is the growing of sufficiently large, perfect crystals. We have found that ideal measurements cannot normally be made with samples below a few mm minimum dimension. However, useful information may be obtained from samples down to a few tenths of a mm. The inaccuracies no doubt arise from internal reflections.

It is a matter of luck whether crystals grow suitable faces through which Raman measurements can be made. If a crystal is easily handled in the air or even in a good inert-atmosphere box, conventional grinding and polishing techniques can be used to provide such faces. Otherwise, one is at present dependent on good fortune to provide acceptable natural faces, and quite often also to ensure that such faces are in good optical contact with the glass wall of a vacuum system. Clearly, specially designed apparatus would remove some of these problems. Often 90° and 180° measurements can be combined to avoid the use of a very poor face.

It is possible to purchase goniometer mounting systems for single-crystal work, and these may be useful for exact measurements on well formed crystals. Generally the chemist will find "string and sealing wax" to be acceptable substitutes! Much useful work can be done on the Cary 81, for example by sticking the crystal to the hemispherical lens with glycerol or, if that is too reactive, Nujol. Similarly, a crystal in a vacuum system can be examined at 180° through the side wall without difficulty. The chances of being able to make successful 90° measurements in such a system are much less.

We cannot here give details of crystal setting, or of the determination of the direction of the axes. X-Ray techniques are a sure method of doing this, but it is often found that a morphological study is sufficient for unambiguous orientation, when the crystal is sufficiently well formed for Raman measurements to be made. It is well worthwhile becoming familiar with the optical and morphological examination of crystals, including the use of the polarizing microscope. It is also necessary to become familiar with the detection of twinning.

4.3.3 *Birefringence*

Refraction phenomena in anisotropic solids are conveniently described in terms of the optical indicatrix. This triaxial ellipsoid (or ellipsoid of rotation for uniaxial materials) describes the refractive index of the material for a given polarization direction of light propagated in it. In general, when the laser beam strikes the surface of a birefringent crystal perpendicularly, it is split into two components polarized at right angles. The directions of the electric vector are those of the major and minor axes of the elliptic section of the indicatrix formed by the plane perpendicular to the propagation direction. If the section is very nearly circular, interference of the two beams may result in effective depolarization of the radiation (Porto *et al.*[113]). Similar considerations apply to scattered Raman light. This description ignores the details of extraordinary refraction, but it is adequate for most circumstances.

This splitting of the exciting radiation must be avoided if meaningful measurements are to be made. Owing to a similar splitting of the Raman radiation it is pointless to analyze between the axes of the appropriate elliptic section. A possible procedure is to use the axes of the indicatrix for the directional scattering parameters. For orthorhombic and higher crystal symmetries these are the crystal axes; for monoclinic crystals the unique (generally y) crystal axis is coincident with one axis of the indicatrix; for triclinic crystals no necessary relationship exists.

Any face of a crystal may be examined provided that the axes of the relevant elliptic section(s) have been previously determined as extinction directions under the polarizing microscope. Similarly it may be necessary to examine an important molecular direction in a crystal which does not coincide with an axis of the indicatrix. Ideally in this case the crystal should be rotated until the relevant elliptic section of the indicatrix has an axis coincident with the required direction. This condition can also, in principle, be determined under the polarizing microscope.

Final adjustment is easily made in the spectrometer, provided that the emerging laser beam is reasonably well defined. The state of polarization of the emerging laser beam can be tested with Polaroid film (the two components usually showing slight lateral separation), and the crystal should be adjusted until the unwanted component is at a minimum. For very thin crystals, where coherent recombination is possible, the direction rather than the state of polarization of the emergent beam should be tested.

An advantage of the Cary 180° collection system is that exciting and Raman light are subject to the same ellipse, so that only one adjustment is necessary. For other systems the collection direction must be separately examined under the polarizing microscope. In the absence of a polarization scrambler, difficulties will arise from variation of grating transmission with polarization direction unless the indicatrix axes are always used. Similarly, such 90° measurements in general involve *non-orthogonal* axes. It is clear that, for compounds of unknown crystal form, examination under the polarizing microscope is essential before single-crystal Raman spectroscopy can be undertaken.

Some inaccuracies are inherent in all systems collecting Raman light of finitely small divergence. The wide-angle collection of the Cary hemispherical and other *f*/1 lenses is disadvantageous because widely diverging scattered light is subject to a section of the indicatrix very different from that along the defining direction. This difficulty is experimentally apparent only in the most accurate determinations and in the case of circular sections of the indicatrix (for measurements made along an optic axis).

In infrared transmission spectroscopy only the *symmetry-determined* indicatrix axes remain constant through absorption bands and over a spectral range. It is therefore preferable to make infrared measurements using such axes. Even so, true absorption maxima may not be observed owing to changes in the amount of *reflected* light. Reflectance data are similarly subject to birefringence-type phenomena, and must also be subject to lengthy analysis before yielding resonant frequencies. In infrared transmission work, it is likely that the recombination of the beams will be coherent, so the sample may constitute a retardation plate.

4.3.4 *Non-centrosymmetric Crystals*

Apart from the complications encountered with modes both Raman- and dipole(infrared)-active in non-centrosymmetric crystals, some of these crystals may rotate the plane of polarization of light. These are in particular the enantiomorphous classes, without planes of symmetry or inversion axes (C_1, C_2, D_2, C_4, D_4, C_3, D_3, C_6, D_6, T, O). However, some other non-centrosymmetric classes (C_s, C_{2v}, S_4, D_{2d}) are predicted to show the phenomenon in some directions only. Clearly, such rotation greatly complicates observation of the Raman spectrum. Specific rotations vary from acceptably low (3°8′/mm for $NaClO_3$) to impossibly high (325°/mm for cinnabar). The Raman spectrum of $NaClO_3$ has been successfully elucidated with an arc source, making due allowance for the rotation.[114] This compound belongs to a cubic crystal class (T), and so the rotation is not complicated by birefringence.

4.3.5 *Choice of Axes for Monoclinic Crystals*

Having considered birefringence phenomena, it is useful to re-examine the monoclinic case, where crystal symmetry determines only one of the optical indicatrix axes. It may be seen that, for *any* face containing this twofold crystal axis, one of the extinction directions lies along that axis. Thus, any such pair of faces at right angles to one another can be used to define a set of three orthogonal axes for determination of the Raman tensor, and for predictions using the oriented gas phase approximation. For example, in the case of Al_2Br_6 crystallographic x and y were used, with z' perpendicular to both of these.[103] However, unless the three axes defined in this way are also the three axes of the optical indicatrix for exciting light of the wavelength used, it is necessary to use both 180° and 90° measurements to obtain all six components of the Raman tensor. This fact appears to have been neglected by Suzuki *et al.*[110] Thus, measurements taken using the third (*ac*) face of their crystal may not be reliable and may contain a mixture of R_{aj} and R_{cj} for both orientations of the analyzer, or of R_{ia} and R_{ic} for both directions of laser polarization.

We consider a cube-shaped monoclinic crystal, with edges *a*, *b*, and *c*, where *b* is the crystal unique or twofold axis in accordance with crystallographic convention. In general, only light propagating perpendicular to the *ab* and *bc* faces retains its polarization with electric vector direction along *a*, *b*, or *c*. Thus, only *a* and *c* may be used as propagation directions. The following measurements are then possible:

90°	*a*(*bb*)*c*	*c*(*bb*)*a*	180°	*a*(*bb*)*a*	*c*(*bb*)*c*
	a(*ba*)*c*	*c*(*ab*)*a*		*a*(*bc*)*a*	*c*(*ba*)*c*
	a(*cb*)*c*	*c*(*bc*)*a*		*a*(*cb*)*a*	*c*(*ab*)*c*
	a(*ca*)*c*	*c*(*ac*)*a*		*a*(*cc*)*a*	*c*(*aa*)*c*

Thus, the (squares of) all six tensor elements can be ascertained, with sufficient overlap between 90° and 180° measurements to determine the change in collection efficiency. The third face could of course be used, provided that its extinction directions were determined. In general this would lead to measurements involving a' and c' rotated from the original *a* and *c* axes. Some 90° measurements would also involve non-orthogonal axes, e.g. $a(ca')b$, and would, on that account, not be useful.

With Al_2Br_6, 90° manipulation of the crystal was impossible owing to the sensitivity of the sample to the atmosphere (including that in an inert-atmosphere box). Assignments therefore had to be made on a limited number of (180°) measurements. This is not ideal where the oriented gas phase approximation is used, but inspection of e.g. the C_{2h} Raman tensors shows that assignment to *crystal* symmetry species is quite satisfactory with such limited information.

4.3.6 *Optic Axis Measurements*

It was noted previously that, for nearly circular sections of the optical indicatrix, the two beams may interfere in a birefringent crystal. Such directions are known as 'optic axes'. The experimental difficulty arises in defining any propagation direction precisely. Thus, Porto *et al.*[113] have shown that it is possible to send an *uncondensed* laser beam along the optic axis of a good crystal with minimal loss of polarization, but Raman light collected in that direction may not be analyzed.

It may be seen that ellipsoids of rotation have one circular section only; hence, crystals with this 'cylindrical' type of symmetry (one greater than twofold axis) are said to be 'uniaxial' (one optic axis) whilst lower symmetry systems are 'biaxial'.

In the uniaxial crystal classes (those belonging to the trigonal, tetragonal, and hexagonal systems) the optic axis is a crystal axis, and as such it is generally used as one of the Raman tensor axes. For routine measurements it is undesirable to have to (a) remove the laser condensing lens, (b) obtain a perfect crystal, and (c) align it so that the laser passes accurately along the optic axis. Since this is the only way in which polarization can be retained for propagation along the optic (three-, four-, or six-fold) axis, we have to consider the alternatives. We again find that it is possible to complete the measurements using both 90° and 180° collection:

90°	$a(ba)b$	180°	$a(bb)a$
	$a(bc)b$		$a(bc)a$
	$a(ca)b$		$a(cb)a$
	$a(cc)b$		$a(cc)a$

and equivalent measurements, where c is the unique axis and $a \equiv b$. The (degenerate) pair $R_{ac}{}^2$ and $R_{bc}{}^2$ are obtained in both ways and serve to calibrate the collection efficiencies, or $R_{cc}{}^2$ may be used. Another method is deliberately to 'depolarize' the laser with a quarter-wave plate (the remaining circular polarization is irrelevant to incoherent collection and detection), and to send it along the optic axis. Collection at 90° analyzed perpendicular to the axis gives a measurement containing $R_{aa}{}^2+R_{ab}{}^2$ which may be disentangled by comparison with $a(ba)b$. Collection at 180° (no analyzer necessary) gives the same result and may be compared with $a(bb)a$. This method may be applied for any choice of a and b axes, the latter being determined by the unique $a(ba)b$ or $a(bb)a$ measurement (see section 4.4).

4.4 Determination of Sign

Since the observed intensity in a Raman scattering experiment is dependent on the square of the appropriate Raman tensor element, it is not directly possible to determine relative signs of tensor elements. Such signs are not meaningful for different crystal modes, but the information may be of interest for independent components both active in a single mode. For example, in many uniaxial crystal symmetries, the combinations $R_{xx}+R_{yy}$ and $R_{xx}-R_{yy}$ belong to different species, and separation of modes of such species depends, effectively, on a determination of the relative sign of R_{xx} and R_{yy}. There is a difficulty of principle in determining this sign directly, because Raman scattering is incoherent. However, it is in general possible to use a different set of axes for the Raman tensor, in terms of which the ambiguity is removed. In the case quoted, new axes at 45°, $x' = x+y$ and $y' = x-y$, give $R_{xx}-R_{yy} = R_{x'y'}$, but $R_{xx}+R_{yy} = R_{x'x'}+R_{y'y'}$. This rotation is possible in a uniaxial material because birefringence phenomena do not distinguish directions in the xy plane. However, some care is necessary in the original choice of x and y since, in general, a third species has tensor component $R_{xy} = R_{x'x'}-R_{y'y'}$. Using this technique, the a_{1g}, b_{1g}, and b_{2g} internal modes of the $PdCl_4{}^{2-}$ and $PtCl_4{}^{2-}$ ions were distinguished[103] in K_2PdCl_4 and $(NH_4)_2PtCl_4$. In these compounds the ion site symmetry is D_{4h}, as for the free ion. Many $M^I_2 M^{II} X_6$ crystals were similarly treated.

It may also be possible to determine relative signs by the requirement of consistency with the oriented gas phase approximation; see $Na_2S_2O_6$, $2H_2O$.[68] This is not usually a very harsh test for the approximation, and so the method may be successful where the approximation is poorly obeyed.

4.5 Work in the Literature

A few specialist workers have in the past performed single-crystal measurements using arc sources, particularly Mathieu and his co-workers. For example, $HgCl_2$,[115] K_2HgCl_4,H_2O,[116] and $LiClO_4,3H_2O$[99] have been examined in this way. In the third case, the data were used to determine the orientation of the water molecules.

Some of the theoretical discussion in the literature has in the past referred to the fluid-oriented phenomenon of a 'depolarization ratio'. Manufacturers particularly still tend to use this term in single-crystal work. There is no excuse for this especially when laser sources are used. Particular geometries isolate particular components of the Raman tensor. One might refer to measurements of particular isotropic (R_{ii}) or anisotropic (R_{ij}) components, but in general it is recommended that the notation of Porto *et al.*[113] be used without further qualification.

Much of the literature has already been summarized under the relevant sections of this chapter. We now consider some of the more interesting work which has not yet been described.

Some chemical interest has been aroused by determination of the factor group fundamentals of inorganic crystalline polymeric materials. Thus, it has been suggested[105] that for many structures $(AB_x)_n$ either B is much lighter than A, or A lies at a symmetry centre and is thereby constrained not to move in Raman active modes. For such materials (where x need not be an integer) the higher-frequency observed fundamentals at least may be described in terms of the vibration of B atoms in a static environment consisting of m atoms A in various BA_m B-coordination polyhedra. Some reasonably consistent group-frequencies for metal oxides have been determined in this way. The coordination polyhedra considered were OM_m where $m = 1$—4. The oxides discussed were TiO_2 (anatase and rutile), SnO_2, ReO_3, WO_3, CeO_2, ThO_2, ZnO, β-Nb_2O_5, V_2O_5, MoO_3, and MoO_3'. References to previous original work on those oxides are in that paper.

Other crystals examined for Raman-active optical phonons by laser Raman spectroscopy include corundum;[117] $LiNbO_3$;[118] $SrTiO_3$;[119]

yttrium aluminium garnet;[120] ZnS;[88] AlN, BN, and BP;[121] benzene;[122] calcite;[113] rutile fluorides;[123] quartz;[124] alkaline-earth tungstates and molybdates.[125] In the case of non-centrosymmetric crystals, Raman measurements are of great use in determining both transverse and longitudinal mode frequencies for Raman- and infrared-active modes. For a summary of the previous literature see Loudon.[86]

A number of more recondite studies have been made using single-crystal Raman spectroscopy. For example, spin waves ('magnons') in the 'antiferromagnetic' materials FeF_2 and MnF_2 have been observed to give rise to Raman scattering.[126,127]

In 'ferroelectric' materials, the electric phenomena are associated with equilibrium deformation along a particular path which also defines a normal mode. This 'soft' mode has a peculiar temperature-dependence,[119] and its Raman scattering is enhanced by the application of an electric field,[128] both references referring to $SrTiO_3$. Many studies conducted by physicists have been concerned with the details of the second-order spectra. In addition, the material $KTaO_3$, having only a second-order Raman spectrum (all the atoms lying at symmetry centres), has been studied.[129] Mitra[130] has studied the multiphonon processes of ZnSe, CdSe, ZnO, and CdS, in both cubic and hexagonal modifications where available.

Birman and Ganguly[131] have reformulated the crystal Raman process in a way which can be used to take account of resonance (Chapter 1). A quantum-mechanical treatment applicable to non-resonant conditions has been given by Loudon.[86] The electronic Raman effect for single crystals is briefly discussed in Chapter 8.

4.6 Bibliography for Chapter 4

Most physics textbooks on the solid state deal with vibrations, though few with much generality. A useful source for advanced theoretical considerations is: T. A. Bak (Ed.), *Phonons and Phonon Interactions*, Benjamin, New York, 1964. This is the report of a conference on the subject.

A useful introduction to crystal symmetry and crystal optics is: C. W. Bunn, *Chemical Crystallography*, Clarendon Press, Oxford, 1961.

Crystal optics are dealt with in more detail in: N. H. Hartshorne and A. Stuart, *Practical Optical Crystallography*, Arnold, London, 1964.

The most useful reference source for all aspects of crystal symmetry is undoubtedly: K. Londsale (Ed.), *International Tables for X-Ray Crystallography*, Kynoch Press, Birmingham, 1959, Vol. I.

Advanced aspects, including Space-group character tables, are given in a collection: R. S. Knox and A. Gold (Eds.), *Symmetry in the Solid State*, Benjamin, New York, 1964.

Linear polymer vibrations are dealt with in a number of texts, for example: R. Zbinden, *Infrared Spectroscopy of High Polymers*, Academic Press, New York, 1964.

A condensed summary of point symmetry group theory, all of which generalizes easily to factor groups, is given in ref. 2, i.e. E. B. Wilson, J. Decius, and P. Cross, *Molecular Vibrations*, McGraw-Hill, New York, 1955. In particular, these authors deal with the correlation theorem, and give both a method and a literature summary for the construction of symmetry coordinates.

In many ways the most lucid account of site symmetry methods is still that of the original author, i.e. ref. 100, R. S. Halford, *J. Chem. Phys.*, **14**, 8 (1946). This account deals in a qualitative way with most aspects, and in particular with the ideas subsequently developed into the oriented gas phase approximation. Note, however, that the $|\boldsymbol{k}| \approx 0$ selection rule was not properly understood at that time, so that, inter alia, rhombic sulphur is discussed in terms of the vibrations of the body-centred unit cell. The primitive cell for rhombic sulphur has only four S_8 molecules.

Some useful review articles are in refs. 102 and 86, i.e., S. S. Mitra, *Vibration Spectra of Solids, Solid State Physics* (Ed. Seitz and Turnbull), Academic Press, New York, 1962, Vol. 13, pp. 1—80; S. S. Mitra and P. J. Gielisse, *Progress in Infrared Spectroscopy* (Ed. Szymanski), Plenum Press, New York, 1963; R. Loudon, *Advances in Physics*, **13**, 423 (1964) [see also errata, *Adv. Phys.*, **14**, 621 (1965)]. The latter deals with transverse and longitudinal phonon phenomena, and gives the Raman tensors and dipole moment vectors for all crystal point symmetries.

A simplified discussion of practical details and applications appears in ref. 103, i.e., I. R. Beattie and T. Gilson, *Proc. Roy. Soc.*, *A***307**, 407 (1968).

A discussion of crystal vibration symmetry and activity is given in: D. F. Hornig, *J. Chem. Phys.*, **16**, 1063 (1948).

A useful general reference work is: J. P. Mathieu, *Spectres de Vibration et Symmétrie*, Hermann, Paris, 1945.

5

Raman spectra of gases

There is a small but useful amount of information in the literature on the high-resolution vibration, vibration–rotation, and pure rotation Raman spectra of gases. The main interest has been confined to the rotation spectra of symmetrical molecules, since these do not show any low-frequency infrared or microwave absorptions. Excellent reviews by Stoicheff[132,133] on the classic measurements made with discharge sources give rotation spectra of many homopolar diatomics, carbon dioxide, acetylene, ethylene, benzene, and other symmetric molecules, and vibration–rotation spectra of a few others.

It is worthwhile to make these observations on less symmetrical systems too because it is frequently the case that more information is available than that taken from absorption measurements. Unfortunately Raman measurements have always suffered from lack of precision, since the linewidth of the source has limited the resolution in the gas phase to $\sim$0·25 cm^{-1}, whereas infrared measurements (particularly by interferometric methods) can be an order of magnitude more precise than this. Further, a resolution of 0·1 Å in a blue excited Raman spectrum requires a spectral resolution of 1 in 250,000 compared with 1 in 10,000 at 10 μ. Microwave observations are, of course, far more accurate but are more applicable (owing to their severe upper limit of frequency) to molecules having high moments of inertia. Before the introduction of lasers, the 'classic' source has been a cluster of large straight Toronto Arcs mounted into a 'light furnace' surrounding a 1—2 m long Raman tube. In general, laser sources provide longer wavelengths than mercury arcs, with the result that the intensity of the Raman lines tends to be low but the dispersion in spectrographs is greater than for the blue region. Fortunately, rotation spectra tend to be relatively intense. Some experimental arrangements using laser sources have been devised by Porto and his co-workers; most of the experimental findings and descriptions of their apparatus are collected in one major paper.[134] Other experimental papers which have appeared will now be reviewed.

5.1 Experimental

The laser produces a narrow fundamental linewidth (He–Ne$\sim$0·05cm^{-1}; Ar$^+$$\sim$0·25 cm^{-1}) compared with the discharge source,* it absorbs far less power, and it is an extremely efficient illuminator of the gaseous sample (particularly if the sample is placed within the laser cavity). Inside the cavity it is easy to obtain 1 W of power (He–Ne) at 6328 Å or 10—25 W (Ar$^+$) at 4880 or 5145 Å transient through the sample. Spectra Physics, and some other manufacturers, make available a 'cavity extension' for their Ar$^+$ lasers, which enables measurements of this type to be carried out easily.

Barrett and Adams[135] have recently argued very convincingly that it is largely unnecessary to put the sample inside the cavity if the geometry of the system matches that of the collection optics. They have shown spectra recorded using a condensed laser beam and a minute sample (less than 10^{-7} cc). Whether the sample be placed inside or outside the laser cavity, a simple straight transparent glass cell will suffice. Use of Brewster-angle windows or anti-reflection coatings within the cavity is normal to reduce losses that tend to extinguish the laser. Some of the simple experimental arrangements devised by Porto and others[31,134,136,137] are given in Figure 5.1. In the (a) arrangement (taken from a 1965 paper[31]) the gas is contained in a straight cell in which perforated baffles are fitted to reduce the collection of unscattered laser light. The effective *illuminated* volume of gas is $\sim$0·6 cc. The slit-shaped source is used to illuminate the spectrograph after being turned through 90° about the axis of view with a Dove prism. With this equipment the authors were able to record an excellent spectrum of methylacetylene close to the exciting line (at 6328 Å), using the following experimental parameters: laser output 20 mW, power through sample $\sim$1 W, source linewidth 0·05 cm^{-1}; spectrograph dispersion 2·0 cm^{-1}/mm from 300 lines/mm grating in 9th order, cylindrical lens before the plate; exposure 58 hr, pressure of methylacetylene $\sim\frac{1}{2}$ atm, $\sim$50 lines appeared to each side of ν.

*The 4358 Å line of mercury from a water-cooled lamp has a central unresolved line of $\sim$0·2 cm^{-1} width contaminated with a series of weaker satellite hyperfine lines out to 1000 millikaysers from the origin ($\sim$0·3 cm^{-1}).[134] Clearly this situation is far from ideal as a Raman source, since Rayleigh scattering limits the approach to the exciting line ($\Delta\nu\approx$0·5—1·0 cm^{-1}) (not a serious limitation!), and the satellite lines will weakly excite Raman lines, 'spreading' the rotational lines to $\sim$0·3 cm^{-1} width and thus limiting resolution. The linewidth of the argon ion laser is a little less than this (but see Appendix 1).

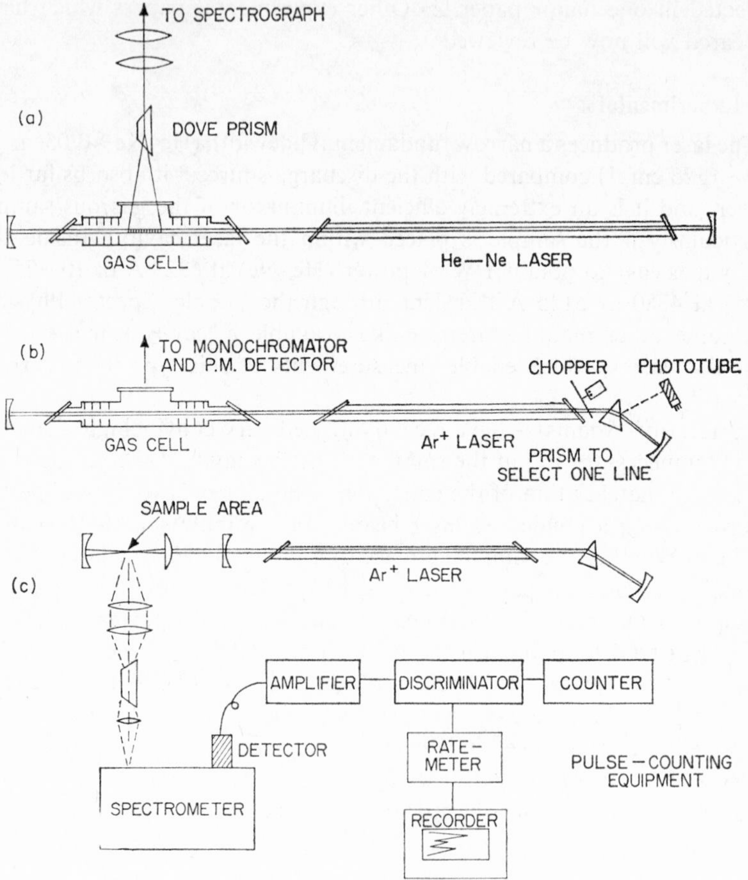

Figure 5.1 Three experimental arrangements for recording gas-phase laser Raman spectra

Arrangement (b) of Figure 5.1 is that of a photoelectric recording instrument powered by an Ar^+ laser. The authors show spectra of nitrogen, oxygen, and carbon dioxide of superb quality recorded by this machine very close to the exciting line.[134] The source was tuned to 5145 Å and had a power of ~5 W; the spectrometer was built around either a Jarrell-Ash 0·5 m single monochromator or a Spex 1400 instrument.

A simple modification of the (b) design was also suggested by the authors so that accurate depolarization measurements could be carried out. It is noteworthy that a polarization scrambler was incorporated immediately before the entrance slit. Also, to limit divergence errors, the beam entering the spectrometer was severely stopped.

Figure 5.1 (c) is taken from results due to Barrett and Adams.[135] These authors describe an apparatus which incorporates photon-counting detection (very clearly explained), the effect of cooling the detector (in their case an E.M.I. 6256S-A with an S-11 photocathode, cooled to around $-40°C$, when the noise is reduced to 1/230 of that at room temperature), and the exact geometry of the focused laser beam resulting from the use of various focusing lenses and mirrors. They were able to record a superb vibration–rotation spectrum of O_2 centred at $\Delta\nu =$ 1556 cm^{-1} (Q-branch) from about 10^{11} molecules contained in a volume of $\sim 10^{-8}$ cc. A signal/noise ratio of better than 50 was obtained in the stronger lines of the O- and S-branches.

Weber and his co-workers[137] have devised gas-cells of greater complexity for use inside and outside the laser cavity. Perhaps the most elegant is that illustrated in Figure 5.2. This cell, which can accommodate as many as

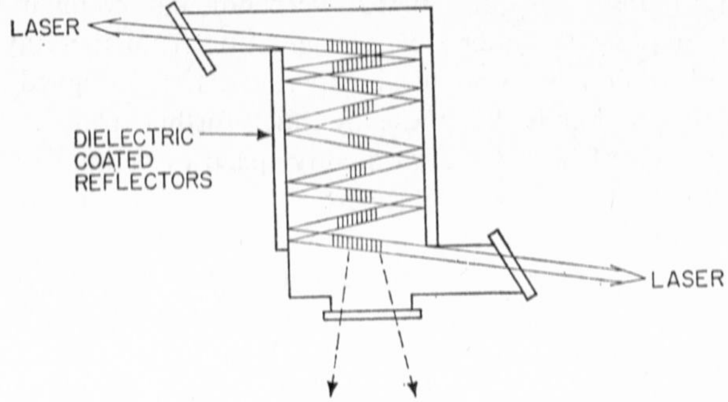

Figure 5.2 A multireflection cell for work inside the laser cavity

eleven traversals within its volume, is designed to be mounted within the laser cavity. The radiation scattered from the volume marked by shading is collected by the spectrograph. An alternative mode of operation is to use a straight cell within the cavity but to focus the beam within the cell. Weber and his co-workers[134] have compared all the configurations, with the following conclusions:

External cells: straight, non-focused beam,	relative efficiency	1
straight, beam focused in sample,	relative efficiency	$\sim$2
straight but incorporating external resonating;	relative efficiency	$\sim$4
straight focused and external resonating,	relative efficiency	$\sim$17
Internal cell: focused in sample,	relative efficiency	$\sim$80

The results are at variance with the conclusions of Barrett and Adams.

5.2 Results

The first point to be made is that almost all the results to appear so far have been recorded before, i.e. they were all determined with discharge sources. There is a considerable improvement in convenience and quality particularly since many spectra can now be recorded photoelectrically. One or two 'new' experiments have been reported but these are discussed in the next section of this chapter.

For rotations, the basic selection rule is that $\Delta J = 0, +2$ and the intensity of the Raman bands is related to the change in polarizability with rotation; i.e. no permanent dipole is required. Thus, the Raman spectrum of a system, such as a diatomic, with a single moment of inertia consists of a series of equidistant lines centred on ν.* (Owing to the low energy of rotational transitions, the Stokes and anti-Stokes spectra have similar intensities.) The depolarization ratio for the Rayleigh line of a gaseous sample is very low (in theory zero for a spherical top) whereas that for a rotational line is the maximum, i.e. 3/4 or 6/7. As a result, it is convenient to observe spectra in the perpendicular configuration, thus severely attenuating the interference originating from the Rayleigh line. This will not, of course, apply when experiments are designed to observe vibration–rotation bands; but these are much further from ν.

By making use of both the high-quality spectra obtainable when using lasers and the fine polarization properties, Weber and his co-workers[134] were able to record ρ-values for the Rayleigh lines of O_2, N_2, and CO_2. The upper limits of these quantities are 0·06, 0·03, and 0·09 respectively.

Weber *et al.*[134] were able to record superb rotational spectra of methylacetylene (with a line separation of 1·14 cm^{-1}), nitrogen (with the rotational lines alternating in intensity owing to nuclear spin[138c]), and oxygen (with the even J lines 'missing' altogether; again see Herzberg[138b]). The same authors also showed a vibration–rotation spectrum† of oxygen with a Q-branch at $\Delta\nu \approx 1556$ cm^{-1} having a width of about 2 Å in the green, and finally they recorded the total scattering cross-section for an individual Raman line (in this case $K = 7$ for O_2) relative to that of the Rayleigh emission; the ratio was 1/420.

In a more recent paper, Barrett and Adams[135] showed superb vibration–

---

*The spectral lines are shifted from ν by $4B(J+3/2)$ where $B = 1/8\pi^2 I_c$, whereas in absorption the frequency of the bands is given by $2B(J+1)$.

†The vibration–rotation spectrum resembles the pure rotation spectrum in having a Q-branch corresponding to the pure vibrational transition with 'satellite' lines [wherein $\Delta J = \pm 2$ and described as S ($\Delta J = +2$) and O ($\Delta J = -2$) branches for a diatomic molecule]. The Q-branches are much more intense than the S or O. More detailed discussions of these phenomena are given by Herzberg.[138a]

rotation spectra of O_2 (complete with the Q-branch distorted towards ν), N_2 (with alternating intensities in the O- and S-branches), and CO_2 (showing ν_1 and three other Q-branches resulting from Fermi resonance; $2\nu_2$, $\nu_1+\nu_2-\nu_2$, $3\nu_2-\nu_2$), and a very large number of vibration–rotation lines between $\Delta\nu = 1200$ and 1460 cm^{-1}, all of which were recorded photoelectrically.

Bernstein has been very active in this field, and at the 9th European Congress on Molecular Spectroscopy[139] he showed some spectra including the vibration–rotation spectrum of methane. The C–H stretching mode was analyzed in detail. In addition, carbon tetrachloride has been examined for the first time in the vapour phase using 5 cm^{-1} slits. It was found that, as expected, the vibrational envelopes observed in the liquid phase 'split' in the vapour. The lowest-frequency line due to the deformation at 218 cm^{-1} is split into a triplet with a prominent Q-branch while the $\nu_1(a_1)$ breathing mode shows two weak satellites at about ± 20 cm^{-1} from its normal vibrational maximum. Bernstein thinks that these may be due to intermolecular effects. The doublet at 762/790 cm^{-1}, due to the asymmetric stretch, loses its doublet character owing to overlap of rotational 'wings'.

At the same conference Barrett and Rigden[140] showed an excellent pure rotation spectrum of N_2O recorded on the gas at $\frac{1}{2}$ atm pressure. We are now able to confirm that, with a 2 W Ar^+ laser and a commercial spectrometer (in particular a Spex instrument), it is possible to reproduce most of the results reported above.

5.3 'New' Experiments

Recently, several 'new' experiments have appeared which demonstrate conclusively that in the near future we can expect a considerable increase in our knowledge of the energy characteristics of gases.

In the first example, a purely conventional approach was used but the new-found convenience of the laser make it of immense potential. Early in 1967 Gasner and Classen[141] studied the molecule xenon hexafluoride, of which the structure was open to discussion. Intuitively one might expect the system to be regular and octahedral but this does not seem to be so. The authors reported spectra recorded using a He–Ne laser on the liquid (at 54 and 92°C) and solid (at 40°C), and were able to interpret these consistently with a tetrameric structure. The vapour density of XeF_6 indicates that the polymeric system breaks on vaporization to monomer units. So convenient has Raman spectroscopy become that the *vibrational* spectrum of the vapour was recorded in this work. The results were 'peculiar', as follows:

Raman: 609 m br (pol.) 520 s br (depol.), 206? cm^{-1}
Infrared: 613 520 cm^{-1}

In an octahedral system, mutual exclusion applies and the activity of bands is expected to be:

Raman: $\nu_1(a_{1g})$ s,sharp; $\nu_2(e_g)$ m; $\nu_5(f_{2g})$ m
Infrared: $\nu_1(f_{3u})$ s; $\nu_4(f_{1u})$ m

Clearly the results are not in agreement with this structure and would indicate an asymmetric array about the rare-gas atom. Theoretical predictions had indicated[142] that this might well be so, but an alternative suggestion must be considered, viz. that of the effect of low-lying electronic levels.

As a further demonstration of potential, the Jarrell-Ash Co.[143] have widely circulated vibrational spectra of a series of compressed gases including CF_2ClCN, $CF_2(CN)_2$, $CCl_2(CN)_2$, CF_2ClCN, and CCl_2FCN. All the spectra were recorded using a single-pass cell, and a signal/noise ratio of ~100 at a slit of 6 cm^{-1} was quite typical for the stronger bands.

In a paper read at Madrid in 1967 Barrett and Rigden[140] described some fascinating experiments on the Raman spectra of discharges. They used an argon ion laser and positioned a specially designed cell within the cavity with the beam focused at the centre of the cell. The cell contained two electrodes across which a discharge could be maintained at the centre of the system. The Raman spectra of the gas under discharge or under its 'normal' environment could thus be recorded. Carbon dioxide was examined close to the exciting line (i.e. in rotation) with a 2900 V 10 mA discharge flowing. Under these conditions a new set of rotational lines was observed, positioned between the normal lines and having about one-quarter of their intensity. The origin of these bands was the subject of considerable discussion at the meeting. It now appears that the lines arise from rotation of the vibrationally excited molecule at 667 cm^{-1} above the ground state. This species is bent, and transitions obeying $\Delta J=1$ are allowed.

In another most unusual experiment, Leonard[144] projected a very powerful nitrogen laser beam (at 3371 Å and of 100 kW power) into the atmosphere and then aimed a Newtonian reflecting telescope to 'look' at the beam 1·2 km ahead of the laser. The scattered light from the air at this point was passed through a variable-filter monochromator and then on to a photomultiplier. Vibrational Raman lines due to both N_2 and O_2

were observed in this experiment. Obviously this unique experiment has immense potential because it must represent the first example of a long-range analysis of a chemical entity. The experiment is truly 'open-ended'. The apparatus used is illustrated diagrammatically in Figure 5.3.

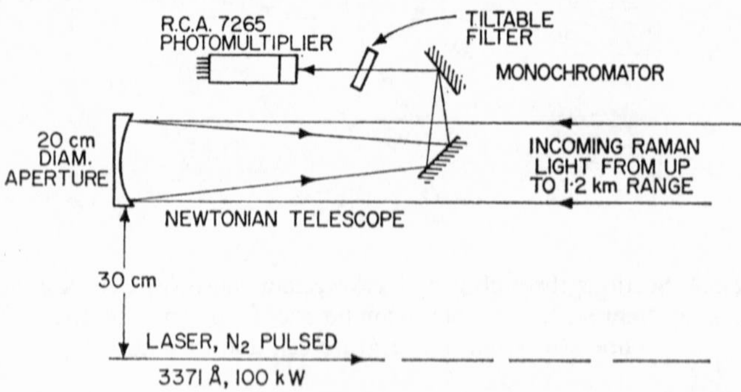

Figure 5.3 The equipment used for observation of Raman spectra at long ranges; the filter monochromator has a resolution of about 35 Å

5.4 Conclusion

Using He–Ne (~50 mW) and Ar^+ (~1 W) lasers and a Spex 140 monochromator and I.T.T.-FW130 detector, at Southampton superb spectra of gases at room temperature and at elevated temperatures have been recorded. By use of this commercial instrument it is easy to record, with suitable cells, pure rotation spectra very close to the exciting line. One such cell, of about 8 cc capacity, has enabled us to approach within 1 cm^{-1} of the exciting line. With the argon laser in an 'external resonator' mode, this commercial spectrometer is sufficiently sensitive to enable spectra of a fairly good scatterer to be recorded at a moderately acceptable signal/noise ratio from a partial pressure of around 100 mm. Spectra have been recorded of O_2, N_2, C_2H_2, and many other gases. Most of the spectra so far measured have been recorded previously, but one is unique—that of chlorine. The line separation in this case is about 2 cm^{-1} but the overlapping of spectra due to the 35 : 35, 37 : 35, and 37 : 37 isotopic species makes it most difficult to resolve the spectral details. For these measurements a special 'cruciform' cell was used (Figure 5.4). The body, of graphite-loaded nylon, was machined to carry four optical flats disposed about an axis and at 90° to one another. The flats seal against soft

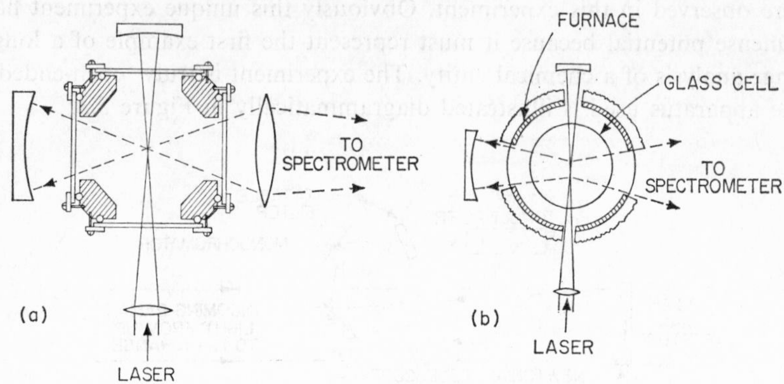

Figure 5.4 Sections through gas-phase Raman cells. (a) Cell designed to reduce stray light so that spectra can be recorded close to the exciting line. (b) High-temperature cell and furnace

rubber rings set in shallow O-ring grooves. By adjusting the clamping screws it is possible carefully to align the flats so that they are parallel to one another. In this way the laser beam is prevented from 'walking' across the cell, impinging on the walls, and increasing the amount of stray light entering the monochromator.

In addition to the study of rotation spectra, a considerable amount of effort has been expended on studying the vibration spectra of vapours at elevated temperatures. It has been possible to record excellent spectra of lead and tin dihalides, the Group III trihalides, and also of elemental selenium and tellurium over wide temperature ranges up to 1000°C. Experimentally these investigations are remarkably easy, and they contrast markedly with the techniques required when arc sources are used. The sample is contained in a simple quartz or hard-glass ampoule (2–3 cm diameter) held in a small tubular furnace. The latter has four small ports cut into it to enable passage of the laser and exit of the scattered radiation. A typical design is shown in Figure 5.4. With halides (other than fluorides) the characteristically strong scattering efficiency results in very high sensitivity, with the result that a partial pressure of only a few cm of mercury is required to record good spectra. Obviously, the absence of optical ('window') material problems, and the sheer convenience of this new method of studying gases and also the appearance in the spectra of bands due to homopolar diatomics, should make laser Raman spectroscopy an important method for high-temperature gas analysis in the future. In one respect it is unique: the spectra are excited from molecules situated within the focus of the laser. Thus, it is possible

Table 5.1 Raman Spectra of Aluminium Tribromide in the Vapour Phase and as a Melt

Vapour (cm^{-1}) 330°C		Vapour (cm^{-1}) 880°C		Assignment* of vapour spectra	Melt* (cm^{-1})†	
59 s	(p)			Dimer	67 wm	ν_4
76 m	(dp)			Dimer	81 wm	ν_7
		93 m	(dp)	Monomer ν_4		
114 w	(dp)			Dimer	114 vw	ν_{12}
139 w—m	(p)			Dimer	140 wm	ν_3
					186 vw	$\nu_1-\nu_2$
203 vs	(p)			Dimer	210 w	ν_2
217 vvw	(?)			Dimer		
230 w	(p)	228 vs	(p)	Monomer ν_1	226 w	
247 vvw	(?)			Dimer		
					284 vvw	$2\nu_3$
					340 vvw	ν_6
		360 vvw	(dp)	Monomer ν_3		
					409 vvw	ν_1
					442 vvw	$2\nu_2$
					489 vw	ν_{11}

*Mode numbering due to Herzberg.
†Data and assignment from I. R. Beattie, T. Gilson, and G. A. Ozin, *J. Chem. Soc.*, *A*, 813 (1968).

to carry out gas analyses in stratified or non-homogeneous systems without sampling, and therefore distorting, any equilibria which may have been established.

Unfortunately, in some cases it has been found that resonance fluorescence spectra are recorded rather than Raman spectra. These may be difficult to interpret. Their characteristic is their high intensity and the regular spacing of band sequences. The 'Raman' spectrum of bromine reported recently by Delhaye[145] is clearly due to resonance fluorescence because the 'first overtone' is almost as intense as the 'fundamental'. For an example of this type of spectroscopy and the analysis of the recorded data, the reader is recommended to consult a paper on iodine vapour.[146]

As an example of an investigation of the vibrational Raman spectrum of a vapour we cite work on aluminium tribromide.[147] In Table 5.1 will be seen spectra of the vapour at 330° and also at 880°C, and also that of the melt. Clearly, at the higher temperature all the dimer units have divided to give monomers whose spectra are in agreement with a planar D_{3h} structure.

At a recent conference on Raman spectroscopy (held at Carleton University, Ottawa, in August 1969), J. J. Barrett read a paper describing interferometric measurements on pure rotation spectra. Since rotation spectra consist of very evenly spaced lines, they make an almost ideal 'source' for a scanning interferometer. By detecting the radiation transmitted by a Fabry-Perot interferometer on a photomultiplier, and scanning the plate separation, Barrett can measure the rotational line separation to very high precision. This method seems to have immense potential in this type of Raman spectroscopy.

It is now convenient, by use of some commercially available spectrometers, and certainly with specially designed systems, to record spectra of gases with ease. As a result it is clear that both infrared and Raman results will become available for vapour samples (e.g. smaller molecules, pyrolysis products, etc.). The analysis of the overall band shapes in the vapour phase arising from the relative intensities of the *O*-, *Q*-, and *S*-branches will also provide much more reliable assignments of the bands to vibrational modes. This approach is familiar in the infrared, and it has been discussed theoretically by Mills for the Raman effect. Also, high-resolution analysis of an increasing number of gases will appear and give us considerably more information than we have at present on interatomic distance values, distortion constants, Coriolis effects, etc. on symmetrical systems of small molecular weight. The Raman spectra of vapours under discharge and at long range will also be of importance.

6

Powders, liquids, and solutions

In Chapter 2 the problem of studying powders and turbid compounds was discussed with particular reference to the introduction of double monochromators in the early 1960s and the subsequent alleviation of many of the experimental difficulties. Before the introduction of the laser, by far the simplest and most satisfactory procedure was to examine compounds either as liquids or, if this were not possible, as solutions. However, problems remained—deeply coloured compounds have always presented difficulties, as have fluorescent or photosensitive species, whilst sample size has also been a severe problem to many organic chemists. Most of the experimental problems have now disappeared, and many new and previously impossible experiments have been reported in which the properties peculiar to the laser have been used to the full.

6.1 Coloured Compounds

In the past considerable efforts have been made to devise special sources which will permit the examination of coloured compounds. The Hg_{5461} green line has found wide application for the study of yellow compounds, but for red, brown, and darker materials emissions from sodium, cadmium, helium, rubidium, etc. are required. Owing to the experimental difficulties inherent in this work, activity has been confined to specialist laboratories of which those of Stammreich (São Paulo, Brazil), Woodward (Oxford), and Lippincott (University of Maryland) are pre-eminent. To give an indication of the success and failure, a few examples of the work are collected in Table 6.1.

The application of lasers to highly coloured materials was proved in 1963 (Daniliheva *et al.*[22]) but the first *chemical* problem was solved in 1965 when Muller and Stockburger[148] examined the Raman spectrum of the thiocarbonate anion. By 1966 results appeared which showed that the introduction of the laser had effected a revolution in this field.[149]

Table 6.1 Examples of Raman Spectra of Coloured Compounds Recorded using Discharge Sources

Ref.	Sample	Exciting line (Å)	Reference and comments
a	Cl_2 (liquid)	5875·6 He	40 min exposure. Lines at 533·4, 535·7, 540·9, 543·1, and 548·4 cm^{-1}, due to $^{37}Cl_2$, $^{35}Cl^{37}Cl$ (1–2 and 0–1), and $^{35}Cl_2$ (1–2 and 0–1) respectively
b	ICl, IBr, I_2, BrCl, and Br_2	Various including 7800·2/7947·6 Rb 7065·2 He	Many spectra of high quality recorded for pure materials and solutions
c	VCl_4	5875·6 He	Assignment proposed
d	$ReCl_6^{2-}$	5461 Hg	Infrared and Raman data reported
	$ReBr_6^{2-}$	5875·6 He	
	$OsCl_6^{2-}$ (solutions)	6678·2 He	
e	CI_4	6678·2 He	Lines at $\Delta\nu = 90$ (*e*), 178 (a_1), 123, and 555 (*f*) cm^{-1}
f	Azobenzene and $Fe(CO)_5$ (solutions)	6438 Cd 8193 K 8195 K r.f. excited	Spectra reported but not assigned. Source described in detail
g	Chromyl chloride (liquid)	6678·2 He 7065·2 He 7281·4 He	Assignment of results given
h	$PdCl_6^{2-}$	6678·2 He	Important discussion of the relative intensities of the bands
	$PtCl_6^{2-}$	5875·6 He	
	$PtBr_6^{2-}$	6678·2 He	
i	CrO_4^{2-}	5875·6 He	Assignment proposed
j	$Cr_2O_7^{2-}$	5875·6 He 6678·2 He	
k	CrO_3Cl^-	5875·6 He 6678·2 He	

a H. Stammreich and R. Forneris, *Spectrochim. Acta*, **17**, 775 (1961).
b H. Stammreich, R. Forneris, and Y. Tavares, *Spectrochim. Acta*, **17**, 1173 (1961).
c M. F. Dove, J. A. Creighton, and L. A. Woodward, *Spectrochim. Acta*, **18**, 270 (1962).
d L. A. Woodward and M. J. Ware, *Spectrochim. Acta*, **20**, 711 (1964).
e H. Stammreich, Y. Tavares, and D. Bassi, *Spectrochim. Acta*, **17**, 661 (1961).
f F. X. Powell, E. R. Lippincott, and D. Steele, *Spectrochim. Acta*, **17**, 880 (1961).
g H. Stammreich, K. Kawai, and Y. Tavares, *Spectrochim. Acta*, **15**, 438 (1959).
h L. A. Woodward and J. A. Creighton, *Spectrochim. Acta*, **17**, 594 (1961).
i H. Stammreich, D. Bassi, and O. Sala, *Spectrochim. Acta*, **12**, 403 (1958).
j H. Stammreich, D. Bassi, O. Sala, and H. Siebert, *Spectrochim. Acta*, **13**, 192 (1958).
k H. Stammreich, O. Sala, and K. Kawai, *Spectrochim. Acta*, **17**, 226 (1961).

Hendra[150,151] reported spectra of a series of square-planar tetrahalo-salts (such as K_2PtCl_4) as powders, using 180° excitation on a Cary spectrometer, and was able to demonstrate that even small samples of brown and black powders usually give excellent spectra. The work was later extended to deeply coloured liquids and solutions, with great success.[152,153] In fact, a spectrum of potassium permanganate (powder) and also of the anion in aqueous solution are to be found in the literature.[154] At Southampton, some excellent spectra of blue and green materials have been easily recorded using the Ar^+ laser.

Therefore most coloured materials no longer present problems for Raman study, except that heating of the sample by the laser can cause thermal decomposition; this is especially serious with dark compounds of poor thermal stability, e.g. some carbonyl compounds, trimethylphosphine–Pt(II) complexes, and $CuCl_2,2H_2O$, but it can frequently be avoided by cooling, by reducing the laser power, or by deliberately diffusing the laser beam before it meets the sample. The enormous power of the Ar^+ laser must be used with particular caution.

There is now a considerable volume of literature in this field but there seems little point in reviewing it since comprehensive surveys have appeared (see Chapter 9). However, it is relevant to list the scope and also the type of information to be obtained from laser Raman spectroscopy of coloured compounds.

6.1.1 *Structure Determinations*

A familiar application of Raman results is in the use of vibrational spectroscopy as a tool for structure determination, and a few examples have been discussed by Ebsworth in a most readable article.[155] Frequently, however, the method has been either limited by the fact that Raman spectra could not be recorded or, where infrared results alone were available, the conclusions tended to be ambiguous. Removal of the experimental limitations has made this rapid, versatile, and non-destructive method more popular.

Bacon *et al.*[156] were able to show, from laser-Raman and infrared results, that the hitherto unstudied molecule hexanitrosobenzene has structure (I).

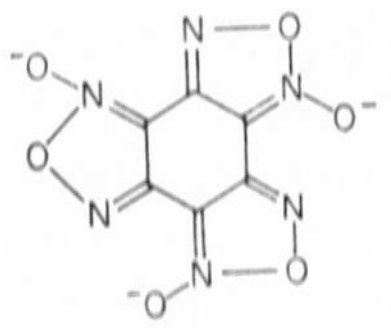

(I)

In one study[153] of charge-transfer complexes, it was possible to find a Raman band characteristic of the I–N charge-transfer bond in the pyridine–iodine system, whilst in another[157] it was shown that in the dimethylchalcogenide–halogen systems a comparison of the far-infrared and Raman spectra led to structures as follows:

$(CH_3)_2S—Br_2(C.T.)—I_2(C.T.)$
$(CH_3)_2Se—Br_2(Cov.)—I_2(C.T.)$
$(CH_3)_2Te—Cl_2(Cov.)—Br_2(Cov.)—I_2(Cov.)$

where C.T. represents a charge-transfer structure of type (II) rather than a covalent structure (III)

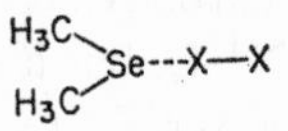

H_3C, H_3C Se---X—X CH_3, :Se, X, X, CH_3

(II) 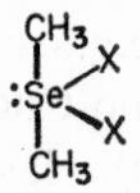   (III)

In another study, the very dark bromo-iodide anion was shown to have bands in both the infrared and Raman spectra, and therefore the ion must be non-centrosymmetric and of C_{2v} symmetry in its caesium salt.[158] The same structure has also been demonstrated, by similar reasoning, for the bromite ion BrO_2^-.[159] There had previously been a tendency to attempt vibrational assignments for deeply coloured molecular systems from infrared data alone, with the result that conclusions have often been of dubious reliability. For centrosymmetric molecules in particular, the infrared spectrum gives very incomplete knowledge of the molecular vibrations; e.g. in the series of square-planar complexes of general formula *trans*-MX_2Y_2, all the symmetric modes are Raman-active, and since mutual exclusion applies they are silent in absorption. Many recent examples of this type of work are concerned with vibrations of the metal–ligand bond. It has been possible in many of these cases to identify reliably the bands due to metal–ligand modes, and to come to conclusions regarding the nature of the bond.[261]

6.1.2 *Fluorescent Compounds*

Reduction in interference from fluorescence is particularly marked in the case of organic compounds such as polymers, aromatics, etc. The polymer field is covered fully in Chapter 7, but it is relevant here to mention some examples of studies made possible by use of lasers.

Table 6.2 Raman Spectra of Nitrogen Trichloride and Related Molecules; Mode Numbering due to Herzberg[6]

	$\nu_4(E)$	$\nu_2(A_1)$	$\nu_3(E)$	$\nu_1(A_1)$
NCl_3	254 m (dp)	347 s (p)	637 vw (dp)	535 s (p)
PCl_3	189 s (dp)	260 ms (p)	494 w (dp)	507 s (p)
$AsCl_3$	155 s (dp)	194 ms (p)	307? ms (370, dp)	412 s (p)
$SbCl_3$	128 s	164 m	356 s (320)	377 s

Results in parentheses are from Kohlrausch[164] and were recorded using Hg arc sources. Other data for PCl_3,$AsCl_3$, and $SbCl_3$ are from Nakamoto.[165]

Perhaps the first successful study of a fluorescent material was that by Perkin-Elmer; as early as 1964 they showed a spectrum of engine oil in which the fluorescence caused relatively trivial interference to the Raman spectrum.

At Southampton a spectrum of the $MnCl_4^{2-}$ ion (tetrahedral) has been recorded by use of a He–Ne laser. The ion is employed in phosphors and is highly fluorescent.

In a general survey by Beattie,[40] are shown some Raman spectra of highly fluorescent compounds. It has been contended by the development department at Spex Industries that in their experience the problem is less severe with the Ar^+ laser than with a red laser but our experience suggests the opposite.

If a sample is left in the laser beam for several minutes before the spectrum is recorded, the fluorescence can be drastically reduced. This 'burning out' procedure has been widely used for some years and does not seem to result in any deterioration of the sample.

6.1.3 *Photosensitive Species*

Some compounds which are insensitive to red radiation decompose rapidly in high-intensity shorter-wavelength light. We first encountered this with di(trifluoromethyl) diselenide.[160] This pale yellow compound, which was available in small quantities (~200 mg), was examined using a Toronto Arc, but selenium was precipitated almost immediately. However, it was easy to record a spectrum using He–Ne excitation, and no decomposition occurred. Other examples include a number of highly labile carbonyl compounds[161] and a series of cyanides of the type X–C≡C–CN (X = Cl, Br, I).[162] All these materials are photosensitive to some degree, but the use of He–Ne sources has been successful.

Perhaps the ultimate in photosensitivity, from the point of view of rate and nature of decomposition, is possessed by the small molecule nitrogen trichloride. Although this compound has been known for well over a century, its unpredictably explosive nature has tended to cool the enthusiasm of most investigators. By analogy with its fluoro-analogue, one would expect it to have a C_{3v} structure, but no serious structural studies seem to have been attempted. By use of the explosion cell shown in Figure 6.1 it

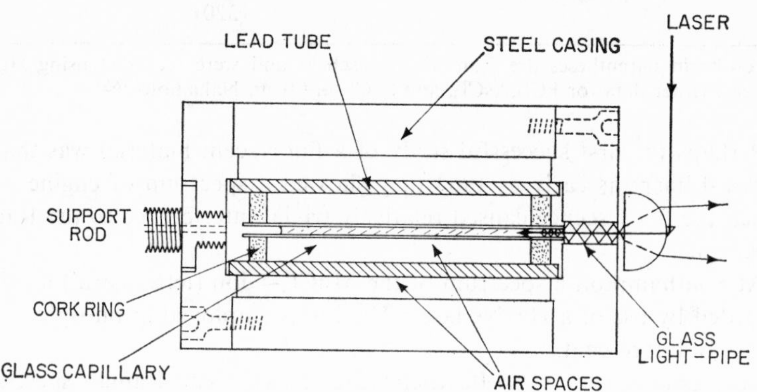

Figure 6.2 Explosion-proof cell for the Cary 81 spectrometer

was possible to record a spectrum of the pure liquid (~5 mg sample) and also of solutions in methylene chloride and cyclohexane.[163] Depolarization data were also obtained and a complete unambiguous assignment was completed on the basis of a C_{3v} structure. The data are collected with those on related molecules in Table 6.2. At no time did the laser cause the compound to decompose. Since the publication of these results two papers have appeared which confirm the data and conclusions.[262, 263]

6.1.4 *Sample Size Reduction*

The dramatic reduction in sample size made possible by using a laser source has been demonstrated in a number of papers. A spectrum of octasulphur (admittedly an excellent scatterer) has been recorded with a sample of only 0·1 mg.[40] In their paper on cyanoacetylenes[142] Klaboe and Klöster-Jensen make the point that the very small specimens available presented no problem because "a conventional Raman spectrophotometer with Toronto Arc would have required sample sizes at least an order of magnitude higher".

This instrumental improvement will assist the spectroscopist and even more the organic chemist. The spectroscopist frequently requires a 'difficult' compound to complete his analysis of a spectrum, or he may require isotopically substituted species of which again only minute samples are available. For example, perdeuterotrimethylphosphine has been prepared, studied in the liquid phase in both the infrared and laser-Raman, and then used to prepare very small samples of coordination compounds which were examined in the Raman instrument.[166,167] Even smaller samples of the arsine were prepared and successfully used to make complexes.

6.1.5 *Highly Reactive Compounds*

Raman spectroscopy has always been popular for the study of highly sensitive species because the sample can be sealed in a glass cell which may be connected or fused to a vacuum line, and even the most reactive species can be manipulated without hydrolysis or disproportionation. The work of Beattie and his co-workers is typical of this approach.[168] In this research, vibrational characteristics of the highly reactive halides of Group V and some transition elements, and also their adducts with simple donors such as trimethylamine, have been determined in detail. Use of laser sources has enabled the size of sample to be reduced, has largely removed the problem of colour, and has made it slightly easier to cool or heat the sample; turbidity is also less of a problem.

One example concerns the coloured labile and hydrolyzable liquid monohalides of sulphur and selenium, X–Y–Y–X (X = Cl or Br; Y = S or Se). The infrared and Raman spectra of these (asymmetric) molecules of C_2 symmetry have been known for some time but they have seemed 'peculiar' in some respects. The Raman spectra of all four molecules have been studied very carefully using laser-Raman and far-infrared techniques.[169] Although the information published on the less reactive molecules was reliable, that on such species as sulphur and selenium monobromides was largely in error because the samples previously used were impure or had disproportionated during examination. Many samples (from various sources) were examined and then repeatedly re-examined during and after purification. With 180° excitation it is not necessary to remove samples during any purification process; the compound can be examined through the wall of the glass vessel. The new techniques are so convenient that more than forty spectra were recorded in less than three weeks. The results of the investigation have been partially confirmed by Bradley *et al.*[170]

Two points transpired from this work. First, the Raman data were much more easily recorded than the absorption results because glass cells are ideal (unlike polyethylene or caesium iodide). Secondly, by using polarizers it was possible to resolve close doublets whereas it was impossible to confirm this in the infrared. The details are given in Figure 6.2 for sulphur monochloride.

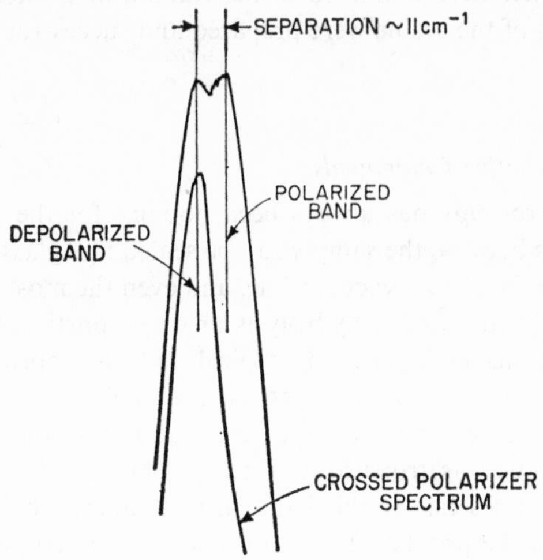

Figure 6.2 Separation of overlapping doublets by use of the polarizer. This example shows the effect very clearly. It is possible to demonstrate the shift on changing the polarizer sense for doublets as close as 3 cm^{-1} to each other

6.1.6 *Specific Experiments*

A considerable number of specific experiments have been reported, and many of them may well be introductions to whole fields in the near future. A few selected examples will be discussed, with particular reference to the potential of the experiments.

6.1.6.1 *Low-frequency vibrations in paraffins.* During 1967 Schaufele and Shimanouchi published a fascinating paper[171] on the very low frequency vibrations of high molecular weight paraffins. Although the bands from these highly turbid compounds are very close to the exciting line, excellent results were recorded for a long series of molecules. Some of the low-

Table 6.3 Some Longitudinal Acoustic Frequencies of Polymethylene Chains C_nH_{2n+2}

n	Order m	ν_{calc}	ν_{obs}	n	Order m	ν_{calc}	ν_{obs}
20	1	116·0	113·6	94	1	26·0	26
	3	327	324		3	75	71
	5	473	475		5	123	121
36	1	65·9	67·4		29	531	536
	3	189	189		31	557	556
	5	306	303	$C_{36}D_{74}$	1	63·0	62·0
	7	403	403		3	177	175
	9	473	475		5	283	281

frequency data are in Table 6.3. The interpretation is particularly interesting because the bands arise from the 'accordion mode' of the chain ⁄\⁄\⁄\⁄ ⟷ /\/\/\/.

Applying the classical 'stretching spring' formulae to the problem:

$$\bar{\nu} = \frac{m}{2L}\sqrt{\frac{E}{\rho}}$$

where $\bar{\nu}$ = frequency of the acoustic mode, m = order of the vibration (number of nodes per chain), L = length of molecule, E = Young's modulus, ρ = density; Raman-active when m is odd.

The results can be carefully analyzed and shown to fit very closely (at least at low orders) to theory. The data yield a Young's modulus of $3{\cdot}58 \pm 0{\cdot}25 \times 10^{12}$ dynes/cm, a value comparable to that determined from X-ray diffraction measurements on polyethylene. This value is much greater than that measured mechanically because the faults in the chains reduce the macroscopic rigidity of the 'crystalline' polymer.

More recently Schaufele[172] extended his measurements to include liquid and frozen low molecular weight normal paraffins (between C_5 and C_{36}). In the solid phase these behave like their higher molecular weight relatives but the liquids show a primary accordion mode which approaches a limiting frequency as the chain length increases. It is considered that this is due to the non-occurrence of *trans*-conformation chains in the liquids when the chain length is less than C_9.

6.1.6.2 *Biological specimens.* Since water is an excellent Raman solvent, it would seem that many biologically interesting aqueous samples could be studied which are inaccessible to infrared absorption measurements.

Before the introduction of the laser the field was not very well represented in the literature.

It is now very simple to record spectra of carbohydrates either as solids or in aqueous solution. From the typical spectra in Figure 6.3 it is clear that the quality is superior to that of infrared spectra owing mainly to the absence of very strong bands which tend to obscure large regions of the spectra. Also, the characteristic Raman linewidth in such systems

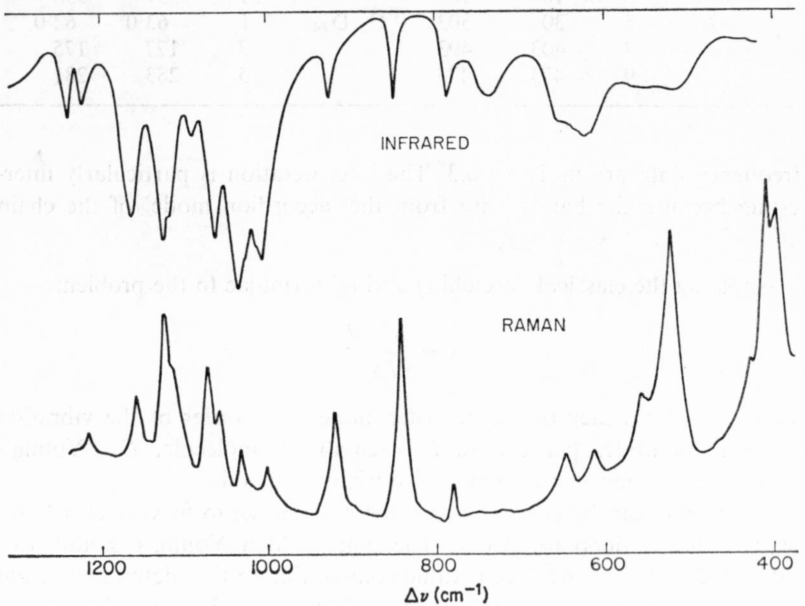

Figure 6.3 Comparison of infrared and Raman spectra for solid glucose

tends to be much less than in absorption. The fact that it is not essential to produce crystals or to dehydrate samples extracted from aqueous solutions is a particularly valuable asset.

During 1967 at the 9th European Congress on Molecular Spectroscopy, Troyanowsky[173] described numerous experiments on organic systems some of which have biological interest, e.g. methyl oleate, glycerol trioleate, and polypropylene glycol. A Kr^+(5681 Å) and an Ar^+ laser (4880 Å) were reported to be superior to the He–Ne device in some respects.

In the spectra of the three crystalline enzymes α-chymotrypsin, lysozyme, and pepsin, Tobin[174] reports that the bands tend to be broad and that the samples are very weak scatterers. The broadness results from the complex vibrational characteristics of these high molecular weight materials and also from hydrogen-bonding, but it was possible to distinguish bands characteristic of a number of structural features in the molecule. Spectra of poly-L-alanine have been reported.[175]

Tobin points out three advantages of Raman spectroscopy over infrared in this type of investigation: (a) the whole spectrum from 3300 to 30 cm^{-1} is obtained in one measurement; (b) the technique is non destructive; (c) the sample size is minute (a few micrograms in his case).

6.1.6.3 *Aqueous solutions.* Using argon ion laser sources, Walrafen[176] has been very active in studying low-frequency modes in liquid water and also in aqueous solutions.[177] An elegant example of this type of study has been produced by Hester and Grossman[178] on the In^{3+}–water system. These authors showed that the viscous nature of these solutions is related to the structure of water molecules around the $In(H_2O)_6{}^{3+}$ ion. They found a peak near 400 cm^{-1} in $In(ClO_4)_3$ solutions which is very intense and increases in intensity as the solution is cooled to -40°C (where it becomes a glass). It is considered that a regular cluster of water molecules, several molecules in thickness, surrounds the complex cation and that the intensity results from the symmetric 'breathing' mode of the large water cluster. In the nitrate and sulphate the anion tends to displace the coordinated water molecules and therefore to 'break up' the structure.

6.1.6.4 *Low and high temperatures.* There are numerous examples of low-temperature studies but they are simply a part of other investigations and therefore comments can be restricted to those made in Chapter 2. Reduction in temperature, apart from producing condensed phases from otherwise normal gases, has in many cases a beneficial effect on the vibrational linewidth, and this can be a tremendous help in resolving multiplets; an example is given in Figure 2.15. Temperature reduction can often help by decreasing the rate of decomposition of an unstable species during examination; phase changes can also be studied conveniently. The manufacturers of the Coderg instrument make available a cell to cool samples to liquid-helium temperature. Some cells have also been designed for the Cary instrument; all these incorporate a 1 cm long by 3—4 mm diameter glass rod to act as a 'light pipe', and a simple cold-finger filled with refrigerant. It is obviously easy to construct an analogous arrangement for the 90° instruments.

Turning to high-temperature spectroscopy, melt spectra have been available for many years and many of the results are collected in a review.[179] A series of papers have appeared describing experiments using laser sources.[180–182] The most up-to-date apparatus is shown in Figure 6.4; by use of this equipment spectra can be recorded up to 1000°C, and

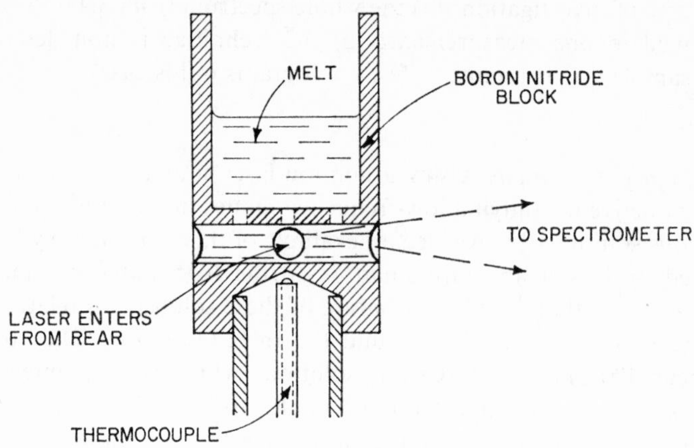

Figure 6.4 A cell for examining the Raman spectra of melts

melts of $NaNO_3$ (350°C), $HgBr_2$ (270°C), and HgI_2 (a deep red liquid) have all been successfully examined. Molten stannous chloride has been shown to contain polymeric $(SnCl_2)_n$ chains with three-coordinate tin atoms. The lead analogue is found to be predominantly the $PbCl_3^-$ ion. Molten cryolite (Na_3AlF_6) has been studied at 1030°C; it was concluded that the AlF_6^{3-} ion disproportionates to AlF_4^- and F^-. Mercuric iodide, which undergoes a colour and phase transition at 126°C has been studied; the low-temperature (red) form was shown to be ionic whereas the yellow form and melt are predominantly molecular.[183]

6.1.6.5 *Band shapes in the liquid phase.* Jennings *et al.*[184] reported very interesting studies of benzene using an argon ion laser at 4880 Å. They observed that the half-linewidths of the vibrational bands in the liquid phase were 3—4 cm^{-1} but underneath these there was a broad weak emission with a half-width of as much as 25 cm^{-1}. These 'wings' are caused by orientation phenomena, and by measuring the width it is possible to compute the orientation relaxation time for the system. By studying seven lines of benzene ($\Delta\nu$ between 200 and 3100 cm^{-1}), and measuring the width of the 'wings' in each case (varying from 3·9 to more than 35 cm^{-1}), the rotational relaxation time was computed; the value $1{\cdot}38 \times 10^{-12}$ sec was similar for each line. The depolarization ratio for both the main line and the orientation wings was $\rho_s \approx \frac{3}{4}$, in excellent agreement with theory. The authors also applied their reasoning to carbon tetrachloride and there seems little reason to believe that the method is not of general application.

6.1.6.6 *Miscellaneous.* Good Raman spectra of glasses have been obtained by using He–Ne (6328 Å, 80 mW) and Ar^+ (5145 Å, 200 mW) lasers. Heat-absorbing glasses tended to fluoresce but gave no trouble. The spectra were interpreted in terms of a structural unit based on a distorted SiO_4 tetrahedron.

In another experiment,[185] mercuric iodide was studied under high pressure, with a diamond anvil cell and 0° excitation, and it was shown that the spectrum is pressure-sensitive. The high-pressure form (above 13 kbar) is identical with the yellow high temperature form. Other experiments in this field are included in ref. 186.

6.1.6.7 *Qualitative analysis.* Raman spectroscopy has always been useful for the identification of some groups difficult to spot by infrared analysis, e.g. C=C, C≡C, C=S, and S–H. Dr. H. Sloane of the Cary Instrument Co. has suggested some further uses. (*a*) It is possible to define the substitution of a benzene ring from Raman band patterns. Combination with i.r. data makes analysis much more reliable than with one technique alone. (*b*) The N–H stretching vibration gives rise to a strong Raman band whilst the OH feature is weak. The reverse is true in absorption. With the two techniques, both groups can be identified in one compound.

6.2 Conclusion

By far the greatest number of laser Raman papers to have appeared have described experiments relevant to this chapter. The laser has effected a revolution in *convenience* as well as in versatility. Some experiments seem to require the peculiar properties of the laser, but recent developments in the use of long-wavelength discharge sources suggest that coloured powders as well as liquids are now accessible.[187] The main improvement is that these new results may now be recorded in a routine manner. As instruments find their way into inorganic and organic/biochemical laboratories, the number of papers reporting full vibrational data (i.e. infrared and Raman) will proliferate. The conclusions must be intrinsically more reliable than those based on infrared measurements alone.

7

Polymers

Polymers are usually very highly stereoregular, and therefore many of their vibrations have no infrared activity. On the other hand, the more symmetrical vibrations are Raman-active. In addition, use of depolarization data is potentially most helpful in assigning the observed spectra to fundamental modes. Although infrared spectroscopy is applied to polymers almost as soon as they have been first synthesized, and is very valuable as an analytical or structural tool, Raman spectroscopy has not been of widespread value owing to experimental problems. Using discharge sources the whole field has been dogged by fluorescence and turbidity problems but the laser has revolutionized the study. Reviews of the subject are available from Schaufele[188] and Hendra.[189]

7.1 Spectra recorded before the Advent of the Laser

The first Raman spectrum of a polymer was recorded as early as 1932 when Signer and Weiler[190] examined specially prepared polystyrene with some success although fluorescence was a major problem. A small number of successful studies have since been made of a number of simple polymers such as polyethylene, polytetrafluoroethylene, and polyoxymethylene. In most cases the data can only be described as poor; signal/noise ratios as low as 5 have been accepted as normal. More recently, some improvements have been effected, and Maklakov and Nikitin[191] have obtained a very good spectrum of polypropylene. An excellent review has been produced by Nielsen,[192] and the data from this are collected together with some more recent results in Table 7.1.

It is apparent that, although some very useful results have been produced, the coverage of the polymer field is sparse. In 1967 Schaufele produced an excellent laser-excited spectrum of isotactic polypropylene recorded photoelectrically, and gave evidence that a revolution was upon us.[193]

Table 7.1 Raman Spectra of Polymers recorded before 1967 using Discharge Sources

Year	Polymer	Authors	Remarks
1932	*Polystyrene*	Signer and Weiler	Specially purified styrene polymerized thermally below 150°C. Absence of oxygen and catalysts essential to avoid a high continuous background. Comparison with spectra of ethylbenzene and styrene led to a partial assignment.
1951	*Polystyrene*	Palm	Poor spectra of carefully purified material. Depolarization ratios obtained. High background a problem.
1957–1961	*Polyethylene*	Nielsen, Woollett, and Holland	Bands observed (cm^{-1}) and suggested assignments: 1065 A_g (skel.); 1132 B_{2g} (skel.); 1171 B_{1g} (CH_2 rock); 1295 B_{3g} (twist); 1419 B_{2g} (CH_2 wag); 1441 A_g (CH_2 def.); 1464; 2722; 2847 (νCH sym.); 2883 (νCH asym.); 2908; 2932.
~1960	*Polytetrafluoroethylene*	Nielsen *et al.*	Observed bands (cm^{-1}) and suggested assignments; 292 (CF_2 twist); 385 (due to amorphous phase); 733 (CF_2 def. sym.); 1226 (νCF_2); 1304; 1382 (νCF_2 sym.). All bands very weak.
1958	*Polyvinyl acetate, polyacrylic acid, and polyethyl acrylate*	Simon, Kunath, and Heintz	Very fine spectra. Partial 'group frequency' type assignment given.
1959	*Polyethylene*	Tobin	Polyethylene spectra better than Nielsen's but no new lines observed.
1959	*Polypropylene*	Tobin	Bands observed at 808, 1155, 1333, 1457, 2831, 2884, 2921, and 2968 cm^{-1}. All lines weak (signal/noise ratio ~5) but following assignments suggested: 808 and 1157 (Me); 1333 (CH); 1457 (asym. Me def.); ~2900 (CH str.).
1959	*Polytetrafluoroethylene*	Tobin	High background; no lines observed.
1961	*Polyoxymethylene*	Tadokoro *et al.*	Bands observed at 1493, 1344, 1229, 1091, 929, and 919 cm^{-1} (signal/noise ~5). Specially purified samples used. Normal coordinate analysis given. Crystalline form assumed to be hexagonal.
1963	*Polyethylene and molten polyethylene*	Brown	Adequate spectra of solid recorded photoelectrically. Melt-depolarization ratios obtained. Structure removed on melting, with considerable changes in spectra.
1963	*Perdeuteropolyethylene*	Brown	Good spectrum recorded on this compound.
1964	*Polypropylene (isotactic)*	Maklakov and Nikitin	Excellent spectrum of polypropylene down to $\Delta\nu = 268$ cm^{-1}. Melt also examined: spectrum similar to that of solid. Spiral structure assumed to decay on melting. No spectral evidence for this.
1965	*Polyoxyethylene and* $[(CH_2)_nO]_\infty$ $(n>2)$	Matsui *et al.*	Adequate spectrum using arc excitation, but results nearer than 800 cm^{-1} to the exciting line diffuse. Preliminary results only.

7.2 Experimental Techniques

The results produced by Schaufele have been recorded using conventional 90° excitation, and with the sample in the form of a rod, or a block (especially when the polymer is transparent and clear), or a pellet drilled with a taper hole when it is opalescent (see Figure 7.1). The results are particularly good close to the exciting line. In fact, superb spectra of translucent solid paraffins have been recorded as close as $\Delta\nu = 20\ \text{cm}^{-1}$ to the exciting line (section 6.1.6.1).

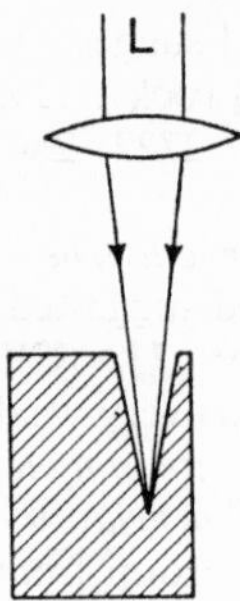

Figure 7.1 The arrangement used by Schaufele to record the spectra of polymers in the form of pellets

A number of spectra have been recorded with the Cary spectrometer. The advantage of this instrument rests in the 180° illumination system, so that no special sampling technique is required. All that is needed is that the sample should be placed in intimate optical contact with the plane surface of the hemispherical collector lens. It has been found that translucent polymers can be examined conveniently as powders in glass ampoules or as bulk specimens by simply pressing them against the lens. Unfortunately, translucent and powder specimens destroy the polarization characteristics of the source, and therefore no depolarization data can be expected in these cases. For clear specimens such as polymethyl methacrylate, it is convenient to study the sample as a rod. If the rod is of appreciable diameter (greater than about 0·2 in) the depolarization data are moderately reliable; if it is smaller, multiple reflections inside the rod will partially remove the polarization characteristics and render the depolarization results unreliable. Frequently, it is necessary to record spectra of sheet samples, and therefore the thickness of sample is of importance. Polymethyl methacrylate samples have been studied with thicknesses of 0·005, 0·010, 0·025, 1/32, 1/16, 1/8, and 1/4 inch, and it was observed that the intensity of the spectrum was improved up to about 1/16 inch. Dispersion of the beam in films of opalescent polymers is a help in that samples thinner than 1/16 inch can be studied usefully.

Some experiments have been made using lasers other than the He–Ne device. By use of the 4880 Å line of the argon ion laser at ~1 W on a solid specimen of polytetrafluoroethylene, an excellent spectrum was observed. The signal/noise ratio on bands out to $\Delta\nu = 800$ cm^{-1} seems to be slightly more favourable than that when using a 70 mW He–Ne device. On the other hand, the signal/noise ratio at $\Delta\nu = 1400$ cm^{-1} may be ~10 times better than for the red laser spectrum. In a way, it is surprising that the improvement is not more marked. The lack cannot be traced to laser instability or other instrumental causes; maybe the turbidity is influential. Fluorescence is always a problem in polymer spectroscopy, and this is a little more severe when using the blue source even in favourable cases such as with polytetrafluoroethylene. Therefore, many compounds which give trouble from fluorescence with the red laser would probably not be examined any more satisfactorily with a blue or green source. With polymer specimens which show fluorescence it is particularly advantageous to leave the specimen in the spectrometer and exposed to the laser for a time ($\frac{1}{2}$–8 hr) before recording spectra. The background signal decays exponentially with time and an improvement of 10× is frequently obtained.

It has been shown that useful orientation data can be obtained by examining the spectra of polymers as fibres or stretched films in a polarized laser beam. Spectra have so far been recorded on the Cary spectrometer with the arrangements shown in Figure 7.2. Considerable

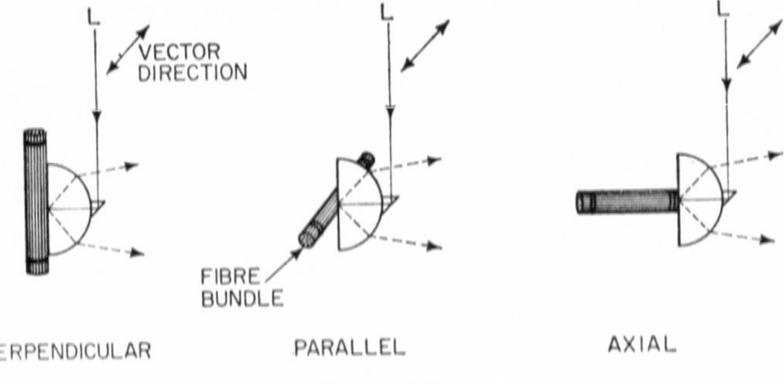

Figure 7.2

orientation effects were observed first in bundles of polypropylene fibre and tape,[194] and this was followed by data on polyethylene[195] and Nylon 66.[196] The significance will be discussed in the following sections where relevant. The data can, of course, be obtained on any instrument, but difficulty may be encountered if 90° illumination is attempted. The necessity to traverse quite considerable amounts of polymer may well spoil the polarization of the laser beam unless the fibres or film are very clear in transmission and sufficiently thick.

Returning to depolarization ratio measurements: we have made numerous attempts to record ρ values on especially prepared polymer specimens whose turbidity has been deliberately suppressed by quenching (e.g. polyethylene and isotactic polypropylene) but with little success. It appears that although seemingly transparent, sufficient discontinuities in index of refraction exist in these cases to cause considerable scrambling of the polarization of the laser and Raman light.

7.3 Raman Spectra of Specific Polymers

7.3.1 *Hydrocarbons*

Polyethylene. The vibrational spectrum of polyethylene has been the subject of a number of papers. The classical Raman results are included in Table 7.2 whilst the infrared bands are at 2919, 2851, 1475, 1170, 1063, and 722 cm^{-1}. The frequencies observed have been fully assigned with some confidence. Thus, in 1963 Shimanouchi stated that almost conclusive

Table 7.2 The Raman Spectrum of Polyethylene*

Major lines	Assignment Schachtsneider and Sneider[2000,201,202]	Amendments Sneider[197,198,272]	'Dichroic' results Hendra and Willis[195]	Boerio and Koenig[208] low temp. data	Evaluation of all data
1063 m	(B_{2g}) skel		$\|=\perp$ (A_g)	1068/1065 (B_{2g})	(B_{2g}
1130 m	(A_g) skel		$\|$ (B_{2g})	(A_g)	(A_g)
1170 w	(B_{1g})CH_2r.	Ditto	$\perp$ (B_{1g})	(B_{1g})	(B_{1g}
					1295 (B_{3g}
1296 ms	(B_{3g})CH_2tw.		$\|=\perp$ (B_{3g})	1295/1297 (B_{3g})	1297 (A_g
1370 vw	—	(B_{2g})CH_2w		(B_{2g})	(B_{2g}
1417 wm	(B_{2g})CH_2w	Resonance $2\times B_{1u}$ $2\times B_{2u}$ and A_g	$\|$ (B_{2g})	(A_g) Correlation doublet.	1417 (B_{3g}
1440 ms	(A_g)δCH_2	(A_g)δCH_2	$\perp$ (A_g)		1440 (A_g
1461 m	—	—	—	—	$2\times B_{1u}$

*Line group descriptions in brackets.

assignments had been made for this molecule.[197] More recently, doubts have been expressed regarding the frequency of the B_{2g} methylene wagging mode, and Sneider has suggested[198] that its value is ~1378 instead of 1415 cm^{-1} as previously suggested. This latter band is thought to be due to an overtone or correlation splitting of the A_g CH_2 deformation (see previous page.).

A laser Raman spectrum was recorded of stretched polyethylene fibres in three orientations as depicted in Figure 7.2. As expected, the first two arrangements gave similar spectra (Figure 7.3) considerably different

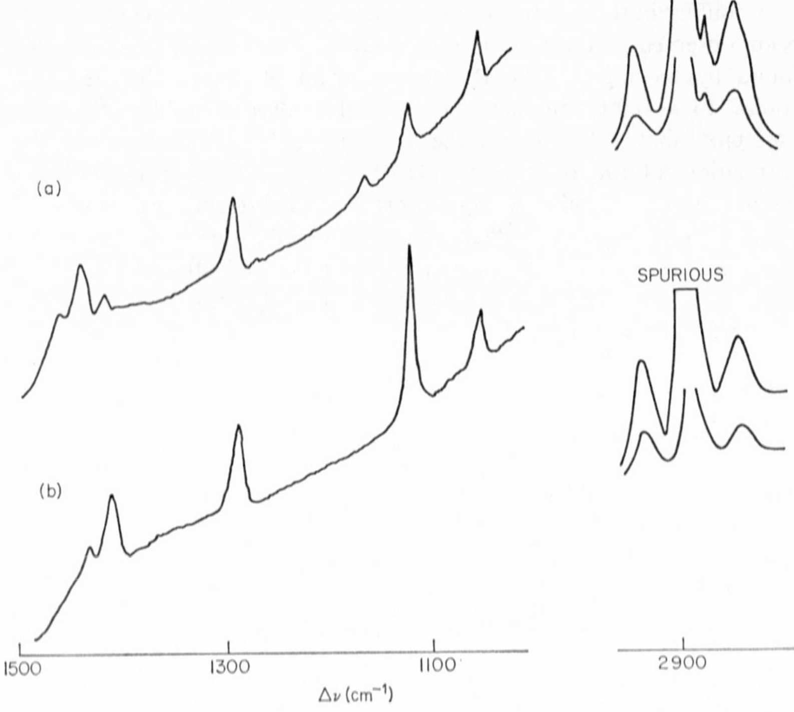

Figure 7.3 Raman spectra (He–Ne laser) of oriented polyethylene fibres: (a) vector perpendicular to fibre axis; (b) vector parallel to fibre axis

from that recorded in the parallel mode. If one makes the somewhat sweeping, but nevertheless eminently reasonable, assumption that $\partial\alpha/\partial q$ is maximized in the same direction as the movement vectors in the normal coordinate, if symmetry permits (cf. section 4.2.5.2) then it is possible to assign the bands as one would do using infrared dichroic data.* It is clear that, although the band at $\Delta\nu = 1065$ cm^{-1} is somewhat non-committal, the band at $\Delta\nu = 1133$ cm^{-1} is clearly 'parallel dichroic'. This strongly indicates that it cannot be due simply to an A_g class mode as originally proposed.[202].

It must be remembered however that the crystal of polyethylene contains two chains per unit cell and that these lie with their planes roughly perpendicular to each other. The Space group, like the Line group of the isolated molecule, is D_{2h}.* The following Line to Space factor group correlations are expected: $a_g \rightarrow a_g + b_{3g}$; $b_{1g} \rightarrow b_{1g} + b_{2g}$; $b_{2g} \rightarrow b_{2g} + b_{1g}$; and $b_{3g} \rightarrow b_{3g} + a_g$, i.e. each chain fundamental is expected to appear as a doublet in the spectrum of the crystalline polymer. The splitting has been observed frequently in the infrared and recently has also been found by Koenig† in the Raman spectrum of a polyethylene sample cooled to -180°C. In fact, Koenig feels that the problematic line at 1415 cm^{-1} may well be due to the A_g δCH_2 mode indicating an enormous correlation splitting of 25 cm^{-1} (1415/1440 cm^{-1}). Bearing these effects in mind it is then possible to re-interpret the 'dichroism' results[195] and to conclude that it is reasonable that the established assignment[202] is correct. A considerable amount of the data published on polyethylene is collected in Table 7.2. A spectrum of polyethylene has also appeared which does not agree with any other![203]

Polypropylene. Considering first the isotactic isomer, a number of Raman spectra have appeared (in papers by Schaufele[193] and others[40,194]). None of these spectra is remarkably better than that due to Maklakov and Nikitin[191] except close to the exciting line. This polymer gives little trouble from fluorescence, and therefore it is possible to record good spectra by use of discharge sources. A spectrum of oriented fibres and films is also available.[194] A spectrum of a sample of unstretched monofilament isotactic polypropylene is given in Figure 7.4.

---

*For a factor group D_{2h} the symmetry limitations on the Raman tensors provide that the following activities apply (adopting the nomenclature of Chapter 4): a_g xx yy zz; b_{1g} xy; b_{2g} xz; b_{3g} yz.

Since the z-axis has been chosen to lie along the molecule (Linn and Koenig),[201] the activities will be a_g and b_{1g} with the electric vector perpendicular to the fibre axis, and a_g, b_{2g}, and b_{3g} with the vector parallel to the axis. Thus it is apparent that both spectra should contain lines due to the A_g modes but only the parallel spectrum will contain the B_{2g} skeletal band. Unfortunately the purity of polarization of the laser is appreciably destroyed by the polymer, and therefore these rules do not *strictly* apply. Also, the symmetry rules do not specify or limit the magnitude of the tensor components, and hence the intensity of the Raman bands.

†F. J. Boerio and J. L. Koenig. *J. Chem. Phys.*, in press (1970).

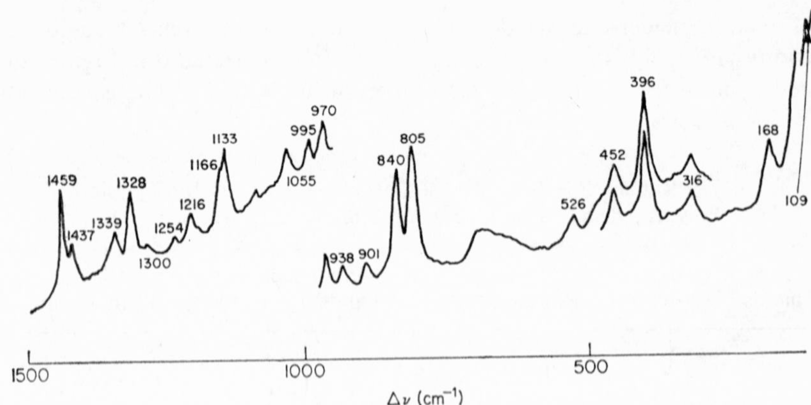

Figure 7.4 Raman spectrum of a single filament of isotactic polypropylene; He–Ne laser, 180° excitation

The infrared spectrum of this polymer shows virtually all the bands that appear in the Raman; further, it will be noticed from Figure 7.4 that many bands appear in doublets of unequal intensity. In the infrared spectrum this also occurs but the intensities are reversed. This strong Raman–weak infrared relationship is frequently encountered and often assists in explaining the origin of spectral features. As a general guide strong R.–weak i.r. bands are due to symmetric modes and the reverse to antisymmetric ones.

In some ways, the syndiotactic species is more interesting because no Raman study of this material has been successful when using discharge sources. A recorded spectrum* agrees tolerably well with the predictions

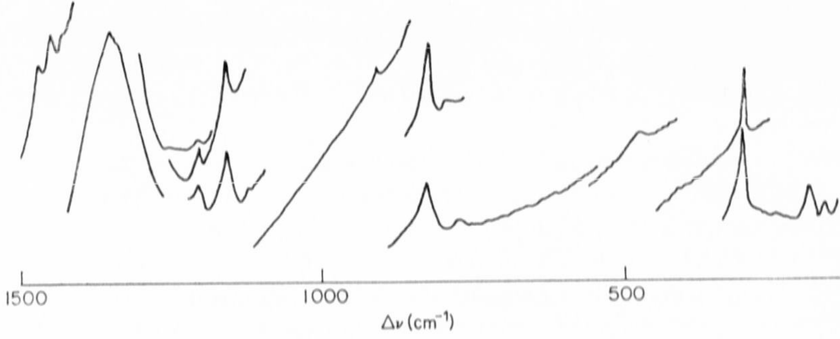

Figure 7.5 Raman spectrum of syndiotactic polypropylene; He–Ne laser, 180° excitation

*P. J. Hendra and G. Zerbi. *J. Mol. Spectr.*, **30**, 159 (1969).

of Schachtsneider and Sneider[205] and Peraldo and Cambini[206] based on coordinate analysis of the system. A spectrum is reproduced in Figure 7.5 and a suggested partial assignment is given in Table 7.3. The calculated

Table 7.3 Raman Spectrum and a Partial Assignment for Syndiotactic Polypropylene; crystalline sample of form I, symmetry D_2

Predicted *A*-class bands[205,206] (cm^{-1})	Observed Raman bands	Predicted *A*-class bands[205,206] (cm^{-1})	Observed Raman bands
2962 }	2960 w	1259	—
2962 }		1191	1202 w
2906	2904 m	1168	1158 m
2882	2892—2872 m br	1041	1040 vw
2856 }	2848 m	996	1000 vw
2855 }		918	—
1463 }	1463 w	839	829 ms
1462 }		542	—
1454 }	1445 mw	347	—
1452 }		315	314 ms
1372	—	199	204 w
1353	1346 vs	176	172 w
1339	1331 vs	66	—

values are based on a D_2 structure [twofold helices with four chains passing through the unit cell] when 26*A* (R), 26B_1 (R+i.r.), 26B_2 (R+i.r.), and 26B_3 (R+i.r.) modes are expected. Very weak intermolecular interaction results in the four bands characteristic of a given line group mode (correlation quadruplet). It has been assumed that all the Raman bands arise primarily from the *a*-class modes, with the result that the predicted values of the frequencies are in remarkable agreement with observation.

Other poly-hydrocarbons. A number of other poly-hydrocarbons have been examined successfully using laser sources, e.g. polybutene, *cis*-polybutadiene, *cis*- and *trans*-polyisoprene, and isotactic polystyrene. Some points are noteworthy.

(*a*) In some of these commercially important samples, fluorescence is not strong enough to eliminate the Raman spectrum. The isotactic polystyrene was specially prepared but commercial atactic samples have given excellent spectra. The polybutene was obtained commercially.

(*b*) In the spectrum of polystyrene most of the features arise from vibrations of the monosubstituted phenyl groups. This is a little surprising since it is frequently stated that the Raman spectrum is dominated by skeletal modes whereas the absorption spectrum is composed largely of bands due to side-group vibrations. A description of the spectrum of atactic polystyrene and of orientation effects therein has appeared.[207]

(*c*) Numerous attempts to record spectra of other poly-hydrocarbons, e.g. *trans*-polybutadiene, have met with limited success owing to marked fluorescence. It is not clear why this should be. Attempts to record spectra of naturally occurring species such as guttapercha, smoked natural rubber, or carbon black loaded vulcanisate have been disappointing for the same reason but in some cases incomplete spectra have been recorded.

Schaufele[208] has recorded spectra of polyisobutylene $[-CH_2C(CH_3)_2-]_n$, and has also been able to record depolarization data since the sample was sufficiently clear for polarization of the laser beam to be retained. Bands

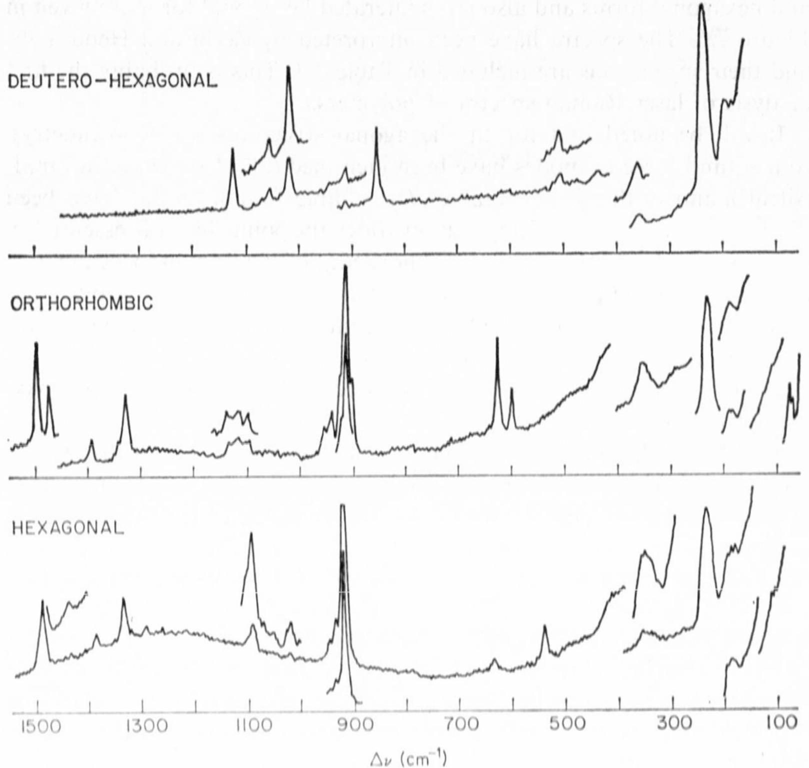

Figure 7.6 Raman spectra of polyoxymethylenes, recorded on powder samples of three forms

at $\Delta\nu = 2919$ and 720 cm^{-1} are clearly polarized and must therefore arise from totally symmetrical modes. Cornell and Koenig* have recorded spectra of *cis*- and *trans*-polyisoprenes and have explained them in terms of structural features in these elastomeric systems.

7.3.2 *Polyethers*

A number of polyethers have been examined successfully. In general, fluorescence seems to be of little importance in these compounds. The spectra have been analyzed in the case of the polyoxymethylenes but not for the more complex molecules.

Polyoxymethylenes. The polyoxymethylenes occur in two crystalline modifications, hexagonal and orthorhombic. In addition, the deuterated species are relatively conveniently available. Spectra of the orthorhombic and hexagonal forms and also the deuterated hexagonal form are given in Figure 7.6. The spectra have been interpreted by Zerbi and Hendra,[204] and their suggestions are included in Table 7.4. This is probably the first analysis of laser Raman spectra of polymers.

It will be noted that for the hexagonal structures (point symmetry), four a_1 and three e_2 modes have been identified (all of these are infrared-'silent'), and in the orthorhombic (D_2) sample four *a* modes have been observed. This example nicely demonstrates the point that it is essential in highly symmetric systems to have a knowledge of the Raman spectrum if an adequate understanding of the vibrational characteristics of a molecule is required. It also bears out the fact that computation of the Raman-active modes is a poor substitute for observation particularly at low frequencies. The calculated frequencies for the *a* modes in orthorhombic polyoxymethylene are 1494, 1319, 958, and 579 cm^{-1} and the observed values are 1491, 1328, 907, and 624 *or* 598 cm^{-1}, showing that errors of up to 10% must be accepted in low-frequency computed frequencies. If our knowledge of the vibrational characteristics advances to the extent that we can compute thermodynamic functions, this error (in the position of low-lying energy levels) will make it essential to *observe* the spectrum.

Other polyethers. Schaufele has reported spectra of a number of other polyethers including polyoxyethylene, a polymer which exists as four helical chains of seven $[(CH_2)_2O]$ groups in two turns per unit cell. A previous study using an arc source (1965) was greatly improved upon. Schaufele and Miyazawa are apparently working on a normal coordinate

---

*S. W. Cornell and J. L. Koenig. *Macromolecules*, **2**, 540, 546 (1963).

Table 7.4 Raman Data for Polyoxymethylenes (Δν, cm⁻¹), and a Suggested[204] Assignment; calculated values in parentheses

Class	Raman band (H form)		Raman band (D form)		Description (replace H by D where appropriate)
			Hexagonal		
a_1	1492	(1491)	1130	(1126)	CH_2 bend
	1338	(1324)	1012	(1031)	CH_2 twist
	923	(962)	852	(838)	COC sym. str.
	539	(499)	508	(470)	OCO bend
a_2	1388	(1419)	1051	(1106)	CH_2 wag
	1095 or 1106	(1110)	969	(974)	COC asym.str.
	—	(928)	845	(813)	CH_2 rock
	235	(223)	200	(205)	torsion
e_1	1439	(1418)	1130	(1131)	CH_2 wag
	1295	(1303)	1085	(1078)	CH_2 twist
	—	(1224)	1064	(1045)	CH_2 rock + COC bend + COC str.
	1095 or 1106	(1077)	908	(896)	COC asym.str. + OCO bend
	936	(916)	845	(806)	COC sym.str.
	635	(649)	—	(632)	OCO bend
	—	(466)	363	(369)	COC bend
e_2	946	(956)	—	(828)	COC asym.str. + CH_2 rock + OCO bend
	—	(525)	493	(486)	OCO bend + COC bend
	190	(191)	180	(187)	torsion
	180	(175)	156	(157)	torsion

Class	Raman band		Description
		Orthorhombic	
a	1491	(1494)	CH_2 bend
	1328	(1319)	CH_2 twist
	907	(958)	COC sym. str.
	624 or 598	(579)	OCO bend + COC bend
b_1	—	(1420)	CH_2 wag
	1096	(1109)	CH_2 rock + COC asym.str.
	895	(932)	COC asym.str. + CH_2 rock + CH wag
	300	(253)	torsion
b_2	1469	(1485)	CH_2 bend
	1114	(1096)	COC asym.str. + OCO bend
	598 or 624	(599)	OCO bend + CH_2 bend
b_3	1388	(1408)	CH_2 wag
	936	(909)	COC sym.str.
	410	(444)	COC bend + CH_2 rock

analysis. An excellent spectrum has been recorded of the planar zig-zag polymer polyoxybutene.[188]

A spectrum has also been given by Schaufele of the high molecular weight molecule $CH_3O(CH_2CH_2O)_4CH_3$. The material is sufficiently pure to enable polarization data to be recorded. The effect of lowering temperature was also noted. Recently we have recorded excellent spectra of hexagonal polymethylene sulphide and selenide at Southampton.

7.3.3 *Polyesters*

Few examples have appeared of Raman spectra of polyesters. However, it is easy to obtain spectra of polymethacrylates, and examples will now be discussed.

Polymethyl methacrylate. The Raman spectra of Imperial Chemical Industries Perspex have been recorded for film, powder, and rod samples. With a $\frac{3}{4}$ inch outside diameter × 6 inch rod, depolarization results were obtained.[209] From these data it was possible to attempt a vibrational assignment based on the group frequency approach. A spectrum has also been recorded of a powder sample of polyethyl methacrylate. Owing to its

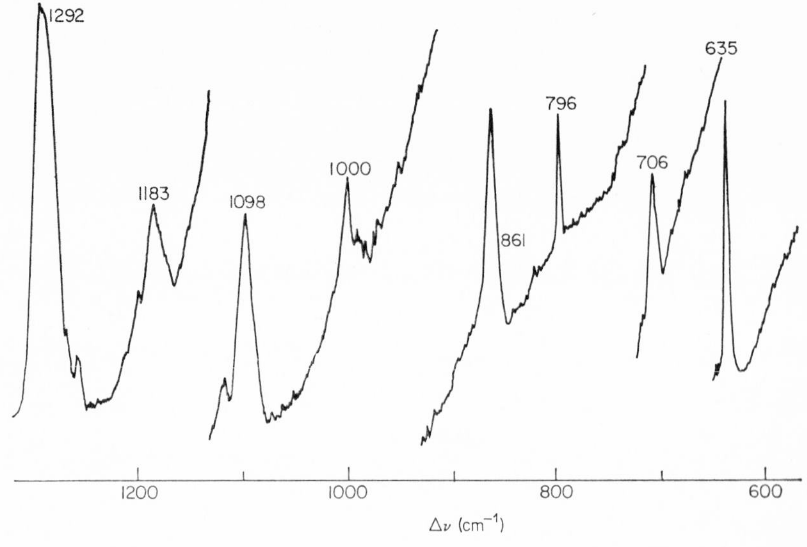

Figure 7.7 Part of the Raman spectrum of a sample of polyethylene terephthalate; note the very high fluorescence, resulting in a steeply sloping background.

complexity and the lack of depolarization data, it was impossible to assign the spectrum to fundamental modes, but it was used in assisting with that of the above methyl compound.[209]

Polyethylene terephthalate. Although this polymer is readily available commercially as powder, film, or fibre, it has been most difficult to record an adequate spectrum free from overpowering fluorescence. Numerous samples have been examined, and one of 0·030 inch thick sheet gave the barely adequate spectrum shown in Figure 7.7.

7.3.4 *Polyamides*

Before the advent of the laser source these compounds were inaccessible, but spectra have now been recorded on many Nylons.

Nylon 66. A spectrum (Figure 7.8) of large-diameter stretched fibres of this polymer has been reported[196] but little attempt has been made to interpret it. It is evident that, as has been observed in absorption,

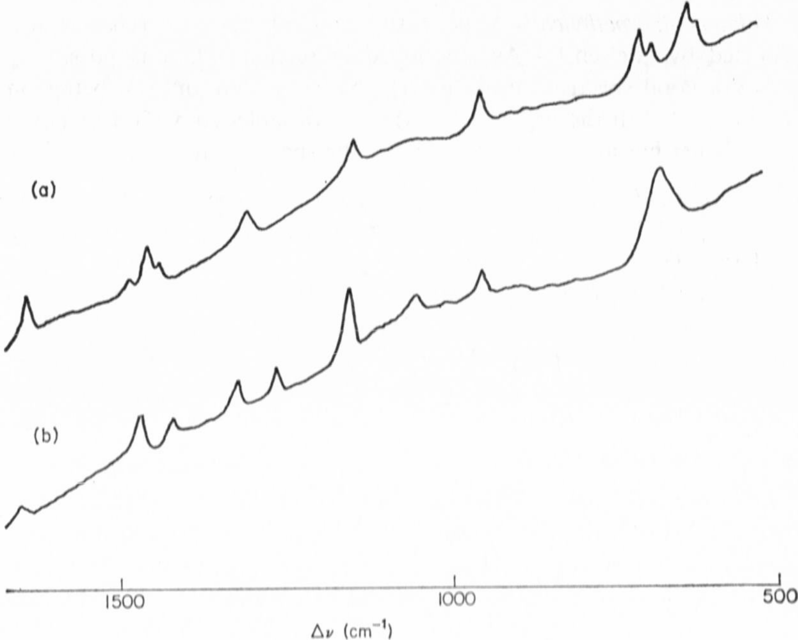

Figure 7.8 Raman spectra (He–Ne laser) of oriented Nylon 66 fibres: (a) vector perpendicular to fibre axis; (b) vector parallel to fibre axis

the Amide I band[210] near 1650 cm^{-1} and considered to be due largely to stretching of the C=O bonds, has perpendicular character. The band at $\Delta\nu = 1264\ cm^{-1}$ behaves similarly to its infrared analogue, being maximized in a parallel direction; presumably this is the Amide III band. As forecast by Miyazawa *et al.*,[210] the Amide II band (near 1550 cm^{-1}) is not apparent in the Raman spectrum.

Some spectra have been recorded at Southampton of various Nylons including 3, 6, 6 : 10, 7, 8, 9/12, 11 and 12. Their quality is a little variable because of fluorescence problems but in general is better than that shown in Figure 7.8. The spectra await analysis but it is immediately apparent that they are dominated by bands due to the $(CH_2)_n$ chains. This is not so in the infrared spectra.

7.3.5 *Polyhalides*

The study of halogenopolymers has been relatively fruitful, and spectra of polyvinyl and polyvinylidene chlorides, polytetrafluoroethylene, and others are relatively easily recorded.

Polytetrafluoroethylene. A poor spectrum of this material has been reported by Nielsen.[192] As mentioned in section 7.2, it is possible to record a good spectrum using the He–Ne laser, and an even better one (Figure 7.9) with the argon ion device. A variable-temperature study of the polymer has also been reported and the spectrum fully assigned.[211]

Polyvinyl chloride. Schaufele[188] has reported a spectrum of this clear polymer and included depolarization details. Bands were observed at 2971 w, 2937 w, 2913 s, and 2847 w cm^{-1} and also at 1437/1430 m, 1323/1308 mw, 1261 w, 1183 mw, 1100 mw, 955 w, 825 m, 691 vs, 634 vvs, 613 ms, 359 m, and 310 w cm^{-1}. Clearly the bands at 691 and 634 cm^{-1} must be due to stretching of the C–Cl bonds, but Schaufele stated that a detailed analysis was in preparation. Recently, Koenig and Druesdow have analyzed their results on PVC. and have concluded that the structure is predominantly syndiotactic.*

Polyvinylidene chloride. This polymer has a distorted planar zig-zag structure with all the CCl_2 groups on the same side of the chain. Some years ago an assignment of the infrared spectrum was attempted by Krimm but the addition of Raman data recently has enabled Hendra, Holliday, and Mackenzie[212] to revise extensively the suggestions. Figure 7.10 shows a spectrum recorded on the powder.

---

*J. L. Koenig and D. Druesdow, *J. Polymer Sci.* A–2, **7**, 1075 (1969).

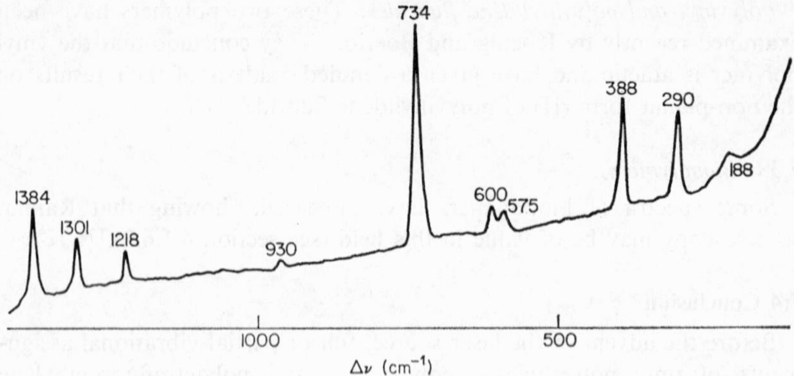

Figure 7.9 Raman spectrum of polytetrafluoroethylene; an argon laser was used

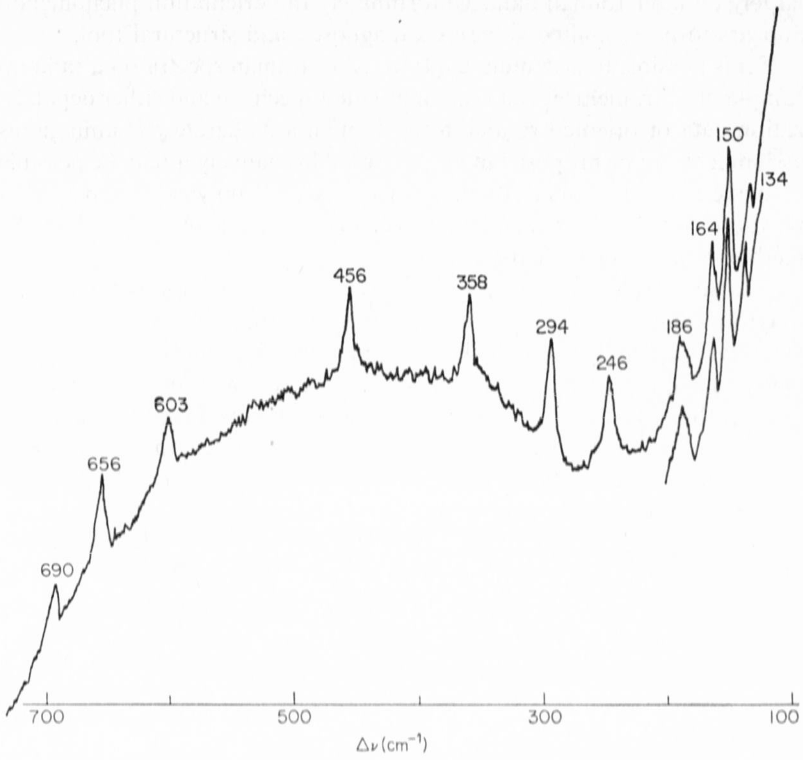

Figure 7.10 Raman spectrum of polyvinylidene chloride (10 mg sample)

Polyvinyl and polyvinylidene fluorides. These two polymers have been examined recently by Koenig and Boerio.* They conclude that the vinyl polymer is atactic and have given a detailed analysis of their results on the non-planar form (II) of polyvinylidene fluoride.

7.3.6 *Biopolymers*

Some spectra of biopolymers have appeared, showing that Raman spectroscopy may be of value in this field (see section 6.1.6.2).[174, 175]*†

7.4 Conclusion

Before the advent of the laser source, full or partial vibrational assignments of only polyethylene, polystyrene, and polytetrafluoroethylene were available based on observation of both infrared and Raman spectra. It is now quite clear that the position has changed completely, since in many of the tables in this chapter new assignments have been given based largely on laser Raman data. Unfortunately the orientation phenomenon appears to be of limited value as a diagnostic and structural tool.

If it is possible to accumulate infrared and Raman spectra on a random sample plus i.r. dichroic data on an oriented specimen and either depolarization data or oriented results in the Raman a moderately unambiguous assignment can be proposed in most cases. Alternatively it may be possible to make a good estimate of the structure of the polymer. A paper on polyacrylonitrile by Angood and Koenig‡ contains full details of the method and is most valuable.

A cursory glance through this chapter might give the impression that *any* polymer can now be examined but this is not so. The list of failures is extensive (e.g. polyacrylonitrile, polyvinyl fluoride, and polyvinylidene fluoride) and is filled with examples of fluorescent materials. Although less serious than it was, this limitation now confronts us again and can probably be eliminated only by use of a longer-wavelength source. Plans to use the quasi-continuous ruby laser (at 6943 Å) are in hand.

Raman spectroscopy enjoys one clear advantage over infrared in the polymer field. If those familiar with the absorption spectra of polytetrafluoroethylene and polyoxymethylene compare their results with Figures 7.6 and 7.9 they will see that in the Raman spectrum the details are not obscured by any very intense and broad features. This annoying problem in absorption rarely seems to arise in Raman spectroscopy.

---

*J. L. Koenig and F. J. Boerio, *J. Polymer Sci.* A–2, **7**, 1489 (1969); and *Macromol. Chem.*, in press (1970).

†J. L. Koenig and P. L. Sutton, *Biopolymers* **8**, 167, (1969).

‡A. C. Angood and J. L. Koenig, *Macromolecules*, in press (1970).

8

Miscellaneous Raman experiments

In this chapter three types of Raman phenomena are included: those arising from conventional vibrational transitions in species adsorbed on to surfaces, those due to electronic transitions, and finally a range of phenomena described collectively as 'non-linear effects'.

8.1 Raman Spectra of Adsorbed Layers

Perhaps the classic method of studying the nature and structure of species adsorbed on to surfaces is that of infrared absorption. So important is the method that two excellent texts have appeared (Little[213] and Hair[214]). The accumulated data are confined to the vibrational spectra of the adsorbate–adsorbent system and the effect of adsorption and/or reaction on that of the adsorbate. During 1962 two reports appeared on Raman spectra of this type of system. The species described were all aromatic and were sorbed on to silica gel. The authors, Karagounis and Issa,[215,216] used powdered gel coated at a density of about one monolayer and pressed into a pellet. The specimens were illuminated with a mercury arc source and the spectra recorded on a Steinheil spectrometer. Their data were not re-examined by other authors, or amplified, presumably because the experimental limitations of their techniques were too severe. Late in 1967 and during 1968 papers by Hendra and Loader[217, 218] reported laser Raman results on adsorbed species. The technique was shown to be simple, convenient, and versatile.

The adsorbent, which should preferably contain only a small proportion of fine particles, is packed loosely into a glass or quartz tube closed by an optical flat. The tube may be either placed in the Raman spectrometer as in Figure 8.1 or fitted to a vacuum line. The design is such that the adsorbate can be cleaned by heating and pumping and then loaded with

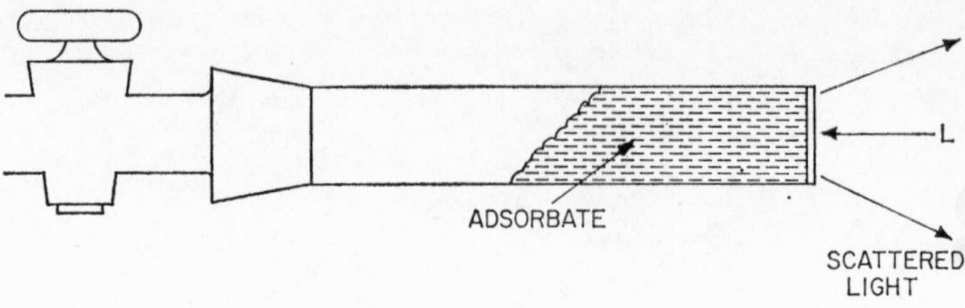

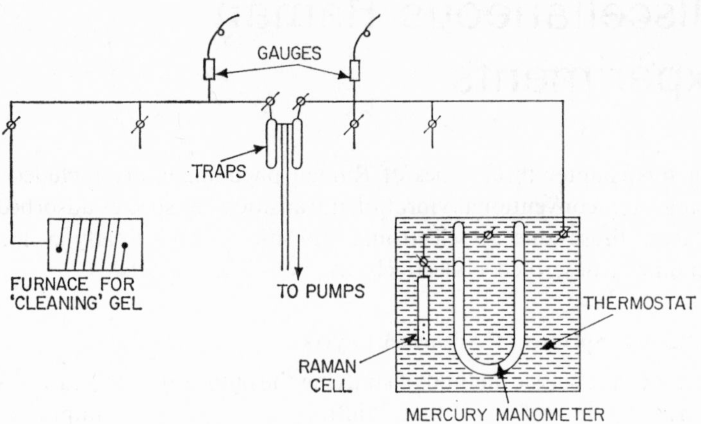

Figure 8.1 Sample tube for examining adsorbed species, and the vacuum system used for studying the macroscopic properties of the gel

adsorbate to various extents before Raman study. If required, the tube is such that Brunauer–Emmet–Teller plots can be completed on the sample *in situ.* Preliminary Raman experiments were carried out on chromatographic grade silica gel. Satisfactory spectra were recorded from carbon tetrachloride, bromine, *trans*-dichloroethylene (Figure 8.2), carbon disulphide, and other species. The following features soon became clear:

(*a*) The method has sensitivity. By use of 70 mW of He–Ne laser radiation in a Cary spectrometer, a band due to bromine can be detected from silica gel of ~170 m^2g^{-1} loaded with ~0·1 monolayer of the halogen.

(*b*) The substrate (in this case silica) seems to have a very weak Raman spectrum and therefore does not interfere markedly with the spectrum of the adsorbate. A prominent band near 200 cm^{-1} due to non-laser emission from the source is the only interfering feature. This property is in marked contrast with infrared results. In absorption, the extinction coefficient and

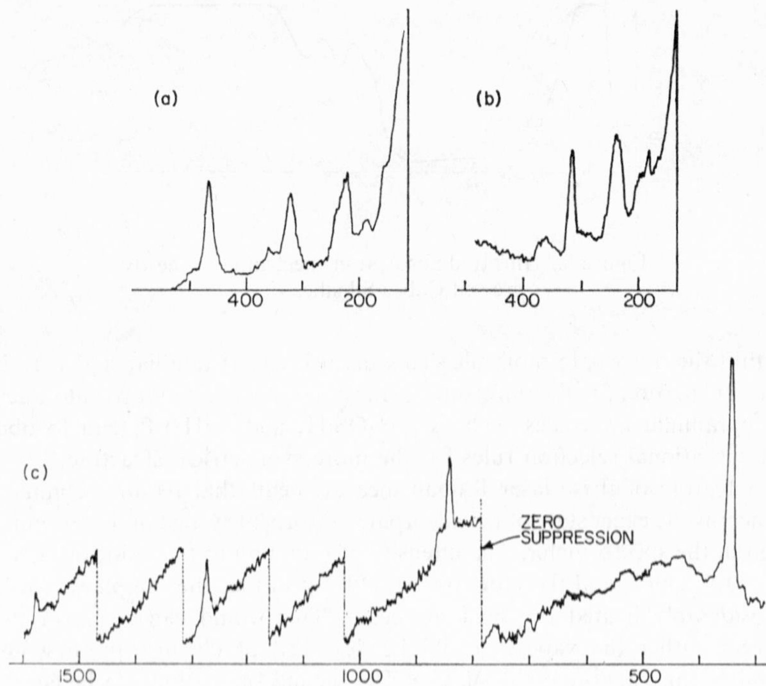

Figure 8.2 Raman spectra of organic molecules adsorbed on silica gel: (a) carbon tetrachloride; (b) bromine; (c) *trans*-dichloroethylene

the high concentration of adsorbent relative to adsorbate in the beam result in obliteration of large parts of the spectrum as far as recording useful information is concerned. An example is given in Figure 8.3.

(*c*) The Raman spectra of the molecules adsorbed on silica gel are almost identical with those of the 'free' molecules in the liquid phase. It was thought that reduction in symmetry resulting from adsorption might make some of the normally Raman 'silent' modes become active but this does not seem to be the case. It is interesting to speculate on why this should be so. In the infrared, adsorption of a molecule can cause activity of symmetrical modes to appear or to increase; e.g. methane on silica shows weak absorption due to ν_1 (the breathing mode) presumably because $\partial\mu/\partial q$ is sensitive to the induction of a small dipole. Intuitively one could expect that asymmetric external fields would not influence the value of $\partial\alpha/\partial q$ to any extent since dipole–dipole or similar interactions would not be expected to change the geometry and the hybridization

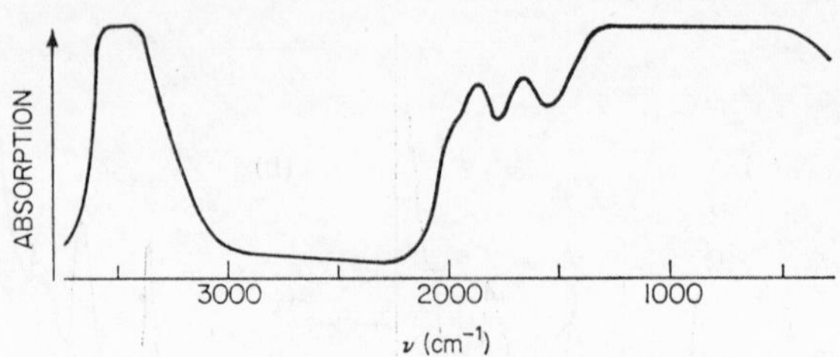

Figure 8.3 Infrared absorption spectrum of a newly pressed Cab-o-Sil silica disc

within the adsorbate molecule. This insensitivity is familiar and may be the explanation for the unfortunate (in some ways) fact that pseudo-linear or pyramidal molecules such as SiH_3OSiH_3 and $(SiH_3)_3P$ tend to obey the vibrational selection rules for the more symmetrical structure.

A feature of these laser Raman measurements that requires comment concerns the exact state of the adsorbate–adsorbent system under examination in the spectrometer. The intensity of radiation at the point-of-view is >200 W cm^{-2}, and therefore one would expect that the sample would be considerably heated by the laser beam. This would cause one of two effects: either the vapour would be desorbed or chemisorption would result from reaction at the surface. The second suggestion can be eliminated immediately from the spectral evidence but the first requires more scrutiny. The intensity of the spectra was found to be insensitive to time, suggesting that little desorption occurred, whereas if serious heating were encountered one would have expected migration of the vapour from the hot to the surrounding cold regions. There is just a possibility that desorption could be rapid and migration sufficiently slow for the sample in fact to be in the gas phase. Multiple reflection of the beam might enhance the intensity of the source to such an extent that the vapour spectrum could be excited. This far-fetched proposal was eliminated by examination of silica gel under 2 atm of nitrogen. No sign could be found of a band due to the $\nu N{=}N$ mode, and therefore it was concluded that, at least for non-absorbent (i.e. white) systems, sample heating is unimportant.

Silica gel is an excellent support. Finely powdered specimens of chromatographic grade gel gave difficulty owing to mobility of the particles beneath the laser beam. This movement, which need be only minute, causes a high degree of 'noise'. Cab-o-Sil specimens of about 325 m^2 g^{-1} were quite satisfactory if firmly tamped into the sample tube. Broken pressed pellets (1 mm thick $\times \sim 25$ mm outside diam. are useful) were also satisfactory, but large molecules do not readily diffuse deeply into the specimens and spurious results are obtained if insufficient time is allowed for equilibration.

Carbon black has been successfully used as a support. A sample of high surface area ($\sim$800 m^2 g^{-1}) loaded with *trans*-dichloroethylene gave good evidence of chemisorbed species. It is considered that sample heating must be inevitable in this case, thus reaction occurs at the surface.

Alon C alumina (50–100 m^2 g^{-1}) was found not to be a good support owing to a tendency to 'colour up' on adsorption.

Preliminary results have also been recorded on industrially important catalyst systems, e.g. unsaturated hydrocarbons over mixed oxides, and carbonyls over silica–alumina. The work can be difficult because the surface area of these supports tends to be limited (typically 3–30 m^2 g^{-1}), and many of them are coloured and therefore weak spectra result. Mixed oxides produce moderately complex 'background' spectra (Figure 8.4 shows two typical examples) but fortunately the bands are narrow and

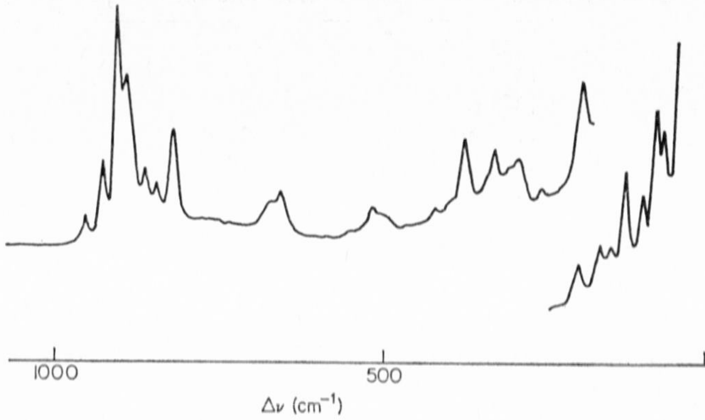

Figure 8.4 Typical Raman spectra of an oxide surface; bands due to adsorbed species are superimposed on this background

often of sufficiently low intensity for features due to adsorbed species to be discerned. Improvements in experimental technique are rapid, with the result that the intensity problem is responding to attention.

Since the spectra of adsorbed species tend to be faint it seems obvious to attempt to use the argon ion laser. This has been attempted at Southampton with various results. The spectrum intensity in favourable cases can be considerably enhanced but absorption of the laser power *must* be avoided if serious heating of the sample is to be avoided. In a very crude experiment, a blackened copper specimen fitted to a chrome–alumel thermocouple in an evacuated jacket was heated to 220°C by an argon ion laser in the Cary spectrometer (~1·5 W at 4880 Å). Clearly this heating is unacceptable. A spectrum of acrolein adsorbed on to Cab-o-Sil at between 1 and 2 monolayers has been recorded successfully using this source.

8.1.1 *Chemisorption*

An example of chemisorption which has been studied in detail is that of acetaldehyde on silica gel.[218] Infrared measurements had shown that the νC=O vibration was not in evidence when the molecule was adsorbed, but little other useful evidence was available owing to obliteration of the 1400—700 cm^{-1} 'fingerprint' region of the spectrum. The Raman data for this system are in Table 8.1. It is clear that the spectrum very closely

Table 8.1 Raman Data for Acetaldehyde Adsorbed on Silica Gel

Adsorbate–SiO_2	Liquid paraldehyde	Adsorbate–SiO_2	Liquid paraldehyde
1459 m	1452 s	839 m	839 s
—	1372 w	—	755 w
—	1345 w	—	575 w
—	1192 m	527 vs	526 vs
1174 m	1169 w	477 s	474 s
—	1101 w	—	372 w
—	945 w	—	274 w
—	857 w		

resembles that of liquid paraldehyde, the cyclic trimer of acetaldehyde, and therefore it is suggested that chemisorption does not occur but that physically adsorbed paraldehyde predominates at the surface of the gel. The ability to observe vibrational spectra in the 'fingerprint' region is clearly invaluable.

Recently, the systems pyridine over SiO_2, SiO_2/Al^{3+}, Al_2O_3, and mordenite have been studied. It is possible to distinguish bands characteristic of liquid, physisorbed monolayer, and chemisorbed pyridine.[264]

In conclusion it is worth emphasizing the advantages of Raman over infrared study of adsorbed species (Refs. 219 and 265 will be useful to researchers.) The information available tends to be more complete since obliteration of large parts of the frequency spectrum is not serious. The geometry of the experiment makes it feasible to study spectra of systems at high or low pressures or temperatures, and even *in situ* in flow systems; i.e. it is possible to investigate heterogeneous catalysts in their normal environments whilst actually operating as catalysts. This latter application has immense potential in commercial manufacturing chemistry.

Some of the conclusions made above have been confirmed in a Russian report of mercury arc excited spectra of sorbed species.[220]

8.2 The Electronic Raman Effect

We shall now deal with electronic transitions in the Raman effect. In principle, there is no reason to believe that bands in the Raman spectrum can arise only from rotational or vibrational transitions. Since one is normally using visible radiation to excite the spectrum, involvement of the process in familiar electronic phenomena usually results in fluorescence, absorption of the radiation, or at least enhancement of the Raman transitions due to the resonance Raman effect. However, there are situations where electronic levels lie at low energy (more typical of vibrations), and it is in these cases ($\nu > \Delta E_{\text{elect.}}/h$) that electronic transitions might be thought to be involved in the Raman effect.

Many of the rare-earth ions contain series of low-lying levels when trapped in a host system. The positions of these energy levels are fairly well understood since low-temperature absorption and fluorescent measurements have been made. In principle, it is impossible to observe absorptions for transitions involving these levels, but inclusion in a matrix or host lattice can give rise to 'mixing' of these levels with excited states involving other electronic configurations, and thus give rise to weak dipole transitions. The Raman effect is, however, allowed in these cases and should therefore be of value. Electronic Raman spectra have been observed by Hougen and Singh (Pr^{3+} in $PrCl_3$) and by Chau (Ce^{3+} in $CaWO_4$) using discharge sources but the technique had practical problems; in particular, readily available discharge sources (Hg 2536 Å and Hg 4358 Å were used) cause fluorescence or absorption rather than scattering. It is fortuitous that the energy levels are so disposed in most of the rare-earth systems that little trouble occurs in this respect when using the helium–neon laser.

During 1966 Koningstein[221] showed that the helium–neon laser could be applied to the problem. Using crystals of rare-earth in garnet hosts (2–5% concentration and 10 mm × 3 mm diameter) illuminated with a powerful laser, viewing with a 1 m Czerny–Turner spectrometer (600 lines/mm grating blazed at 6500 Å) and detecting with an E.M.I.-9558B photomultiplier fitted with a solenoid to suppress noise,* Koningstein

*This is an alternative to cooling. A field of ~500 gauss is used. Details in booklet form can be obtained from the manufacturers of the photomultiplier, E.M.I. Ltd., Hayes, Middlesex, England.

has shown spectra recorded from Eu^{3+}, Yb^{3+}, and Nd^{3+} each in yttrium gallium garnet (YGG).[222] In addition, Koningstein and Mortensen have examined the polarized electronic Raman spectrum of Eu^{3+} in YVO_4 crystals and have obtained a spectrum of Pr^{3+} in yttrium aluminium garnet (YAG).

By careful comparison of all the room-temperature garnet spectra followed by careful observation of the effect of lowering the temperature to 77°K, it is possible to distinguish a few weak lines *not* due to lattice modes. To eliminate fluorescence a search is made on the anti-Stokes side of the exciting line since a feature analogous to the Stokes one should appear for all the Raman spectrum. Lines definitely assigned to electronic transitions were observed as follows:

Yb^{3+}YGG	550 cm^{-1}	mw	All except one line depolarized
Eu^{3+}YGG	308	vw	
	345	vw shoulder (clear using perpendicular polarization)	
	388	mw (polarized)	
Nd^{3+}YGG	86	clear in Stokes and anti-Stokes	
Pr^{3+}YAG	604	w	

Koningstein interpreted his data, as far as intensity was concerned, in terms of transitions described by the matrix elements $\langle (4f)^n | \mu | (4f)^{n-1}5d \rangle$. Using single-crystal Raman polarization results, and knowing the lattice modes beforehand, Koningstein and Mortensen were able to determine the orientation of their yttrium vanadate–Eu^{3+} crystal. The Eu^{3+} ions occupy D_{2d} sites. A series of Raman measurements in a variety of polarization senses were made. A line at 404·5 cm^{-1} was not found in the spectrum of Nd^{3+} in YVO_4, and is therefore assigned to an electronic transition. The polarization behaviour of the line was determined and was found to be that of a feature due to a process of *e*-class symmetry. Some other lines were also observed. Comparison with the fluorescence work of Brecker *et al.*[223] leads to the following conclusions:

Assignment	Results of Brecker *et al.*		Raman data	Difference
$^7F_1$	A_2	333·7 cm^{-1}	360 cm^{-1}	26
	E	375·6	405	29
$^5D_0$	A_1	17,183·2	17,202·5	19·3
$^7F_2$	A_1	985·4	995	10
	B_1	1116·1	1067	−50
	B_2	936·4	970	33
	E	1038·7	1018	−20

The above 'differences' are essentially caused by argument as to the position of the $^5D_0$ level. Koningstein and Mortensen consider that their suggestions are correct and point out that Brecker *et al.* may have made no index of refraction corrections in converting wavelength to frequency.[224]

In addition to their practical activities, Koningstein and Mortensen have been most active in discussing the theory of electronic Raman and Rayleigh scattering. The general expressions for the scattering tensor for electronic Raman effects have been derived. Vibronic coupling has been shown to be unimportant, in contrast to the situation for vibrational spectra.[225]

In another paper the same authors have shown that the absolute cross-sections for Rayleigh and electronic Raman scattering of rare-earth ions can be estimated and are $\sim 10^{-3}$ of the values for liquid benzene. The Rayleigh to Raman intensity ratio is shown to be a function of the electronic structure of the lanthanide ions.[226] For Tm^{3+} the ratio is $\sim 10^5$ but for Tb^{3+} and Dy^{3+} it is ~ 30. In a further paper, calculations of the scattering matrix for electronic transitions for the Ce^{3+} ion in a hexagonal field are given and discussed.[227] It also appears that the scattering tensor is asymmetric for some electronic transitions,[266] and also that magnetic fields can induce a Zeeman effect in the spectrum.

It is obvious that laser Raman spectroscopy provides a powerful and convenient method for measuring the positions of low-lying electronic energy levels. This is potentially of value to those interested in solid-state properties and potential laser-active materials. There is no reason to think that activity in the field should be limited to the solid phase, and liquid-laser rare-earth ion solutions should be accessible to this new technique.

8.3 Non-linear Effects

As outlined in Chapter 1, a number of effects related to the Raman effect have been discovered by using laser sources with higher power. These include stimulated Raman spectra, Raman absorption, and phenomena arising from the finite-value nature of the hyperpolarizability tensor. We shall now discuss the experimental findings relevant to these phenomena.

8.3.1 *Stimulated Raman Spectra*

This process, which arises when extremely powerful sources are used to illuminate a Raman-active sample, was demonstrated accidentally in 1962 by Woodbury and Ng[228] who were using a Kerr cell containing nitrobenzene inside the cavity of a ruby laser with a view to concentrating the energy of the 'spikes' into one 'giant' pulse. When they succeeded, they found a new spurious line in the laser output separated from the conventional laser line by some 1345 cm^{-1} (laser at 6943 Å, new line at 7670 Å) the frequency of an intense Raman emission of nitrobenzene. They named the new effect the 'stimulated Raman effect'. Since 1962 an enormous amount of practical and theoretical effort has been put into the study of the phenomenon. The origin is now moderately well understood but its properties are such that it must be described as a laboratory oddity as far as its value to chemistry is concerned. An excellent comprehensive review of non-linear scattering phenomena, and particularly stimulated Raman scattering, has appeared (Schrotter[229]) and the remainder of this chapter is heavily based on this article. For other reviews see Chapter 9.

8.3.1.1 *Properties of stimulated Raman emission.* There are a number of properties peculiar to this process.

(1) The emission consists of ν (the 'giant pulse' laser frequency) + a series of satellite lines on the Stokes and anti-Stokes sides of this. The lines are at shift frequencies which are *strict multiples* of a very limited number of fundamentals (usually only one or two); i.e. if the value of ν were 12,000 cm^{-1} and the only active fundamental were at 1000 cm^{-1}, we would see lines at, say, 9000, 10,000 11,000 13,000, 14,000, etc. cm^{-1}, in addition to ν. See Table 8.2 for examples. The phenomenon appears in all three phases including powdered solids.[230, 267–270]

Table 8.2 Stimulated Raman Data for Bromoform (exciting radiation, ruby 6934 Å)
Active fundamentals: C–Br_3 sym. def. at 221 cm^{-1} (X); C–Br stretch at 531 cm^{-1} (Y)

$\Delta\nu$ (cm^{-1})	Identification	$\Delta\nu$ (cm^{-1})	Identification
221	X	882	4X
311	Y−X	977	Y+2X
442	2X	1105	5X
531	Y	1198	Y+3X
665	3X	1328	6X
751	Y+X		

(2) There is a distinct threshold below which emission does not seem to occur in most cases. It is frequently found that the origin of the threshold is the self-focusing of the laser beam in the sample, arising from the non-constant value of the index of refraction in high electric fields.[231] The self-focusing causes a dramatic increase of flux density through the

Raman-active medium, with a consequent steep rise in the intensity of the stimulated process. When no self-focusing is possible, e.g. in acetone or hydrogen at low pressure, the 'threshold' is not apparent.[232,233] In more normal circumstances (e.g. for liquids) it is found that the threshold is very dependent on concentration[234] and is also affected by temperature.[235]

Estimates of the strengths of the phenomena have been made. At low laser powers the conventional Raman spectrum is apparent and increases in intensity linearly with laser intensity, i.e. the usual $I_{\text{Raman}} \propto I_{\text{v}}$ relationship applies. At the 'threshold' the selected Raman emissions increase in intensity by $\sim 10^4$—10^6 and then increase further exponentially as the laser power rises;[236,237] i.e. the Raman lines in the stimulated effect have intensities similar to that of the *source* line.

(3) The linewidths of the stimulated lines are extremely narrow if a source laser operating in a single mode is used.[230] In fact the lines are considerably narrower than they are in the spontaneous Raman effect. If a source laser operating in the longitudinal mode is used, the linewidth can be very great (typically ~ 100 cm^{-1}).[238,239]

(4) The process makes use of a very limited number of possible Raman transitions (usually only one or two). It has been stated[240] that only transitions giving rise to polarized lines are 'active', but this has been contested.[241] The 'activity' in these cases is related to the state of polarization of the source radiation. The Raman transitions involved are usually vibrational but the effect has been demonstrated in rotation, e.g. the work of Minck on heavy hydrogen.[268] Rotational lines are apparent in the normal mode of excitation, but vibrational lines appear if the laser is linearly polarized. A circularly polarized source produces a spectrum showing that anisotropic scattering has been enhanced and therefore that the rotational lines have increased in intensity. In addition to vibrational and rotational phenomena, stimulated electronic Raman spectra have been observed.[242–244]

(5) The phenomenon has distinct directional properties. In the arrangement: giant pulsed laser→sample→output, the anti-Stokes fundamental at $\Delta\nu_{\text{R}}$ plus the 'multiple' lines at $n\Delta\nu_{\text{R}}$ are concentrated into cones in the forward direction centred about the direction of excitation.[271]

8.3.1.2 *Sources suitable for stimulated Raman.* Since a very intense source (~ 20 MW/cm^2 flux density) is required, a relatively limited number of lasers have been used as sources. The 'classic' device is the Q-spoiled ruby which is described in Appendix 1. Like its smaller relative, this device operates at 6943 Å. Hundreds of compounds have been irradiated with this device but the following other lasers have also been successfully used.

(a) *Nd^{3+} in glass laser; Q-spoiled.* Benzene, carbon disulphide, perdeuterobenzene, and bromoform have been studied. The resulting spectra were frequency-doubled (moving them into the orange) to make them easier to detect.[245]

(b) *Second harmonic of Nd^{3+} laser.* Cyclohexane was studied with this device operating at 5300 Å. A line at $\Delta\nu = 2852\ cm^{-1}$ was observed, as is normal with the ruby, but the line at $\Delta\nu = 801\ cm^{-1}$ was also found.[246] Attempts to observe a second harmonic spectrum in the ultraviolet did not succeed because two-photon absorption occurred.

(c) *Second harmonic of the ruby laser.* Although not widely applied as a source, the near-ultraviolet emission of the frequency-doubled ruby has been used to irradiate benzene. The $\Delta\nu = 992\ cm^{-1}$ line of this molecule has been seen in the 3000 Å area of the near ultraviolet.[247]

8.3.1.3 *Applications of the stimulated Raman effect.* The main application of the effect is when it is used simply as a new source of radiation at hitherto inaccessible frequencies. Unfortunately, although the effect is intense it can be used only over a very short period ($\sim$1 nanosec). These sources can be used to excite the inverse Raman effect and also for other experiments. By use of frequency-doubling plus stimulation, a set of ultraviolet emissions are available.

The short duration of the stimulated Raman effect can be very useful. In a typical experiment on carbon disulphide, a source pulse of $1{\cdot}4 \times 10^6$ W over 12×10^{-9} sec was used. Backward scattering (i.e. at 180° to the excitation direction) amounted to 3×10^6 W but was concentrated into only $0{\cdot}03 \times 10^{-9}$ sec in the first Stokes line.[248] We therefore have a source of intense radiation in extremely short pulses. A related experiment on hydrogen was used to evaluate the lifetime of the excited vibrational states.[249]

8.3.2 *The Hyper-Raman Effect*

As outlined in Chapter 1, this phenomenon is observed when a powerful beam at a frequency ν is passed into the sample and then the scattered light is examined near 2ν. A spectrum somewhat analogous to the conventional Raman spectrum occurs around this 'central' band. The requirement is that $\partial\beta/\partial q$ should be non-zero for a vibration to be hyper-Raman-active, and this symmetry limitation is not the same as that on the $\partial\alpha/\partial q$ parameter. As a result, many of the infrared- and Raman-active modes are hyper-active but also some of the 'silent' vibrations are expected to give lines, and some non-active rotations will give lines (i.e. spherical-top molecules like methane).

The theory of this most fascinating phenomenon has been the subject of a number of papers[251–253] including one which describes very thoroughly the hyperpolarizability of fundamental modes. Such activity is indicated in the group character tables of Appendix 2. The experimental data in this field are very sparse probably because of the dominant experimental problem of low intensity. Terhune *et al.*[254] first observed the effect in water and quartz near the second harmonic of the ruby. In the former case they recorded bands at $\Delta\nu \approx 800$, 1600, and 3300 cm^{-1}, i.e. close to the infrared absorption maxima. In quartz, bands near 450, 800, and 1200 cm^{-1} were found, again analogous to the infrared absorption data. The authors estimated that the hyper-scattered spectrum had a strength $\sim 10^{-13}$ that of the source, and that the intensity at 2ν is proportional to the square of the source intensity. In another report, Maker described the pure rotational spectrum of the symmetric-top methane.[255] In order to enhance the intensity of the effect he was forced to use a pressure of 100 atm in the gas (thus eliminating any fine structure). The spectrum produced showed wings to either side of 2ν but with the emission maximum offset by an appreciable frequency from exactly 2ν. The author offers a suggestion for the origin of this phenomenon.

8.3.3 *The Inverse Raman Effect*

This effect was forecast and observed by Stoicheff and Jones.[230,256] The experiment consisted of illuminating some benzene with intense Q-spoiled ruby radiation at frequency ν, and at the same time providing a continuum of radiation ($\nu \pm 992$ cm^{-1}) some of which could be *absorbed* by passage through a second sample (toluene). The laser emits in the longitudinal mode, so the emission from the benzene is broad (particularly in the first anti-Stokes line). The specimen of toluene absorbs the radiation at $\nu \pm \nu_R$, where ν_R represents *any* Raman-active transition, and the spectrograph records the information. In the inverse effect the absorption to the anti-Stokes is greater than that in the Stokes. The reason is immediately obvious from consideration of the energy level diagrams. The Boltzmann

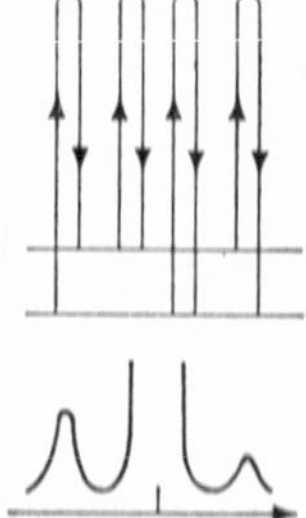

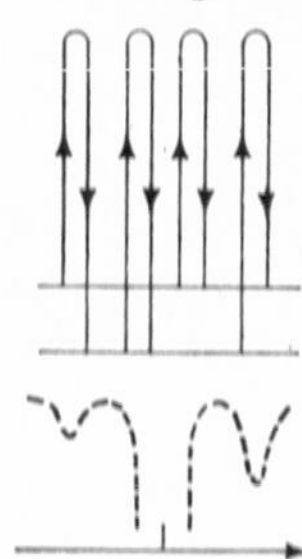

population of the excited vibrational state tends to subdue anti-Stokes emission or Stokes absorption.

In another fascinating report, Macquillan and Stoicheff[257] examined benzene using the toluene emission as a source. The latter has limited usefulness because it has a continuum over only a limited frequency range, $\nu + 900\ cm^{-1}$—$1050\ cm^{-1}$. In a study of benzene they showed a strong absorption at 992 cm^{-1} with a shoulder at 984 cm^{-1} due to the small proportion of $^{13}C^{12}C_5H_6$ present naturally in the specimen. They also point out that the frequency value in this process *does not* exactly coincide with that observed using conventional Raman excitation. The difference is not great but is significant, ~0·2—1·0 cm^{-1}, and they suggest that it may arise from a process related to the Stark effect. Further, they showed that the efficiency of absorption, expressed as a percentage, is linearly related to the power of the source. The lower limit of power for the process to be readily studied seems to be $\sim 5 \times 10^7$ W cm^{-2}, which is not too high to preclude the use of gas lasers in the foreseeable future.

The main problem with a pulsed laser is that of obtaining a continuum which is of sufficient intensity to be photographed, and which also has the same or shorter duration than the laser pulse *and* is synchronized with it. One solution to the continuum problem is to use the incandescence excited in compressed gases by the passage of the laser beam on its way to the specimen. Using this approach Dumartin *et al.* have successfully recorded spectra for benzene, chloroform, and carbon disulphide.[258] Not only were these authors able to observe absorptions on the stronger anti-Stokes side of the emission line, but they were also able to find them in the red-shifted spectrum. For chloroform they observed anti-Stokes lines at $\Delta\nu = 668$, 1216, and 1389 cm^{-1}, and a Stokes line at $\Delta\nu = 668\ cm^{-1}$.

One obvious question concerns the power of the laser beam at ν. What is the result if this approaches the intensity required for stimulated Raman emission in the specimen itself? It is found that 'stimulated Raman absorption' occurs; i.e. very intense absorption sets in. This process has been observed by Dumartin.

8.3.4 *Conclusion*

The theory of the non-linear effects is relatively well understood and has been discussed by Buckingham.[259] However, the experimental observations have tended to disappoint chemists. The stimulated effect is largely useless but there is potential in the Raman absorption and hyper-Raman effects. Raman absorption has potential in two main ways: (*a*) The process is extremely fast ($\sim 10^{-8}$ sec) but is no faster fundamentally than when the Q-spoiled laser below stimulation threshold is used to excite conventional Raman spectra. (*b*) Beer's law applies with all its attendant advantages over the Raman emission processes. Sample pathlength and concentration variations are, of course, of value in this effect. The technique has always suffered from the lack of a really suitable white-light source. Recently, B. P. Stoicheff (at Raman Spectroscopy Conference, Carleton University, Ottawa, 1969) announced that superb results can be obtained by using a Q-spoiled ruby and frequency-doubling a few per cent of the radiation. The red and UV components are then separated prismatically, and the short-wavelength radiation is beamed into a dye such as Rhodamine G. This fluoresces for $\sim 10^{-9}$ sec and acts as the white source for the experiment. Stoicheff has recorded spectra over a wide range of cm^{-1} shifts, and has also shown that depolarization applies in absorption.

The hyper-effect is of great academic potential; it should be possible to measure 'forbidden' mode frequencies, e.g. torsional modes, out-of-phase 'twists' in planar molecules, etc., but it is probably easier to observe bands due to these modes as their harmonics in the conventional Raman spectrum. Recently, P. D. Maker (Raman Spectroscopy Conf., Ottawa, 1969) showed hyper-Raman spectra of water and also of crystalline ammonium chloride. In the latter, 'twisting' of the ion in the crystalline lattice has been observed at $\Delta\upsilon = 380\ cm^{-1}$.

It is a pity that both of these exciting developments are experimentally very difficult to observe.

One cannot close this section without mentioning another inelastic scattering phenomenon, in which the laser acts as the scatterer! Bartell and Thompson[260] devised a system whereby a stream of monochromatic electrons intersected the intense stream of photons inside a laser cavity. The photons acted like a diffraction grating to the electrons, i.e. the photons and electrons exchange momentum but not energy because the photons exist in standing waves. The result is that the electron beam is scattered. This process is, of course, the Compton effect.

9

Sources of information

The purpose of this book is to demonstrate the new-found versatility of Raman spectroscopy resulting from the introduction of laser sources. An attempt has been made to survey the range of experiments already carried out, with a view to stimulating further interest in the subject. The book is not intended to be a review of the literature, although from some aspects the coverage is fairly complete.

General Reviews

J. A. Koningstein, in *Raman Spectroscopy* (Ed. H. Szymanski), Plenum Press, New York, 1967. A very early account and therefore of little current use.

J. Brandmüller, *Naturwiss.*, **21**, 293 (1967). Fairly complete but very concise review of the situation up to January 1967. Excluded non-linear phenomena.

P. J. Hendra and P. M. Stratton, *Chem. Rev.*, **69**, 325 (1969). Fairly comprehensive up to January 1968. Coverage of non-linear effects negligible.

P. J. Hendra, in vol. 1 of a series on vibrational spectroscopy: *Far Infrared Spectroscopy* (Ed. J. Durig), Dekker, New York, 1970. Comprehensive to January 1969 and includes coverage of single-crystal and non-linear effects.

Survey Articles

A. Weber, a series of brief review articles have appeared, of which the most important is in *Spex Speaker* **11** (1966): "Raman Spectroscopy Revisited".

I. R. Beattie, *Chem. in Britain*, **6**, 347 (1967). A survey with much useful experimental information.

P. J. Hendra and C. J. Vear, *Analyst*, April (1970). A survey intended for industrial chemists and analysts.

Other Reviews

Single-crystal Spectroscopy. No reviews available, but see bibliography at the end of Chapter 4. A review of i.r. techniques is: W. Vedder and D. F. Hornig, *Adv. Spectr.*, **2,** 189 (1961); this contains references to 1959, including Raman work.

Gases. Two reviews contain extensive information.

I. R. Beattie and G. A. Ozin, *Spex Speaker*, Feb. 1970.

Polymers. R. F. Schaufele, *Trans. N.Y. Acad. Sci.*, **30** (1967).

P. J. Hendra, *Adv. Polymer Sci.*, **6**, 151 (1969); and an article in *Polymer Spectroscopy* (ed. D. Hummel), Carl Hauser Verlag, Munich, 1970.

Non-linear Effects. W. H. Schrotter, *Naturwiss.*, **54**, 607 (1967). A comprehensive but very brief and cryptic review of the field. A more complete review: N. Bloembergen, *Amer. J. Phys.*, **35**, 989 (1967).

Amongst other sources of information it is worthwhile citing:

Spex Speaker. Often contains useful practical and experimental details on, or relevant to, Raman spectroscopy. (Apply to Spex Industries, Metuchen, N.J., U.S.A.)

Ramalogs. A short periodical confined to laser Raman spectroscopy. Also available from Spex Industries.

Raman Newsletter. Published under the auspices of the National Bureau of Standards, and available free on request to Dr. P. R. Wakeling, 1500 Massachusetts Avenue, N.W. Room 24-C, Washington, D.C. 20005, U.S.A. This publication is devoted almost entirely to laser Raman spectroscopy and is of immense importance because of the very short time-lag between submission of information and its circulation. In addition, this periodical runs a bibliography.

Regarding journals, it must be pointed out to physicists that much work will be found reported in *Spectrochim. Acta, J. Chem. Soc.* (Section A), *J. Amer. Chem. Soc.*, etc., whilst chemists should be careful to scan *Phys. Rev.* and *Phys. Rev. Letters* in addition to other general physical journals.

Books and Articles

Kohlrausch, *Ramanspektren*, Becker and Erler, Leipzig, 1943. A superb account of the subject before the 1939—1945 war. An enormous amount of valuable information is collected in this book but unfortunately it is out-of-print and difficult to buy.

J. Brandmüller and H. Moser, *Einfuhrung in die Ramanspektroskopie*, Steinkopff Verlag, Darmstadt, 1962. An excellent account of the developments in vibrational and rotational spectroscopy since Kohlrausch's work.

E. B. Wilson, J. C. Decius, and P. C. Cross, *Molecular Vibrations*, McGraw-Hill, New York, 1955. A comprehensive account of the theory of vibrations.

G. Herzberg, *Spectra of Diatomic Molecules*, 1950 revised, and *Infrared and Raman Spectra of Polyatomic Molecules*, 1945; both Van Nostrand, New York. In some respects, although rather 'old', the spectroscopist's bibles.

R. N. Jones, a series of biennial reviews in *Analyt. Chem.*; latest edition, **38**, 393 R (1966).

H. Siebert, *Anwendungen der Schwingungspektroskopie in der Anorganischen Chemie*, Springer, Berlin, 1966. A very valuable review of vibrational data on inorganic and coordination compounds.

K. Nakamoto, *Infrared Spectra of Inorganic and Coordination Compounds*, Wiley, New York, 1963. Contains some information on Raman results.

D. Adams, *Metal-Ligand and Related Vibrations*, Arnold, London, 1967. A review of the subject.

H. Szymanski (Ed.), *Raman Spectroscopy*, Plenum Press, New York, 1967. A series of first class articles by prominent Raman spectroscopists including L. A. Woodward.

Appendix 1
The Laser

The term LASER stands for Light Amplification by Stimulated Emission of Radiation, and is used to cover a wide range of instruments producing radiation in the ultraviolet, visible, and near- and far-infrared. All these involve a source or 'pumping' system used to excite a medium to an excited state with a long lifetime. The excitation may be in one or more energy jumps, but in all cases a metastable state is achieved in which the proportion of active species in the device that are in an excited state is greater than that predicted by the Boltzmann relationship. This condition is described as a 'population inversion'. One peculiar property of this state of affairs can be described as follows: when one active species reverts to a lower energy level (and subsequently to the ground state) a quantum of energy is produced which promptly 'stimulates' or induces other excited species to do the same; i.e. as the radiation proceeds through the 'population inverted' material it is amplified in intensity. If this system is placed inside an optical resonator, we have a system analogous to the electronic oscillator; i.e. rays propagated in such a way as to be appropriate to the resonator will pass and re-pass through the excited medium and therefore be amplified again and again. The first such system was the ruby laser described in 1960 by Maiman. In this device a ruby crystal (consisting essentially of Cr^{3+} ions in a crystalline alumina host lattice) is the active medium, with the energy level system given in Figure A1.1. Excitation is achieved by exposing a ruby rod to very intense visible

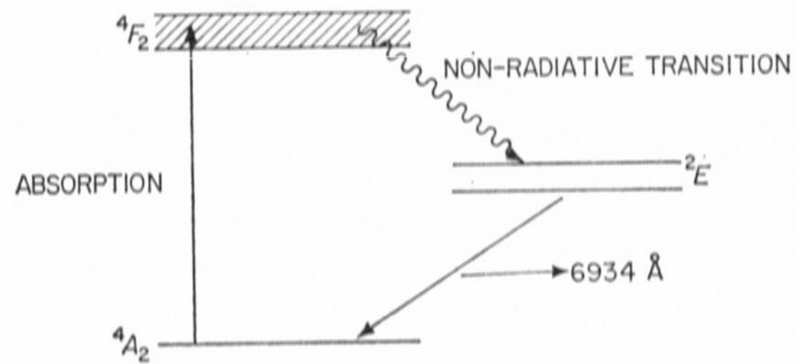

Figure A1.1 The energy levels involved in the ruby laser

radiation, e.g. that from a xenon flashlamp, whilst the optical resonator cavity is produced by grinding the ends of the rod flat, polishing them parallel to one another, and coating them with dielectric multilayer reflector coatings. In this system rays exactly normal to the end-reflectors proceed up and down the ruby rod and are amplified again and again. In addition, they are all 'in-phase', i.e. the wavefronts are in step with one another, thus generating effectively a standing optical wave system inside the device. Rays that do not satisfy these criteria either 'walk off' to the side of the ruby rod or are inefficiently amplified. A typical ruby laser system is shown in Figure A1.2.

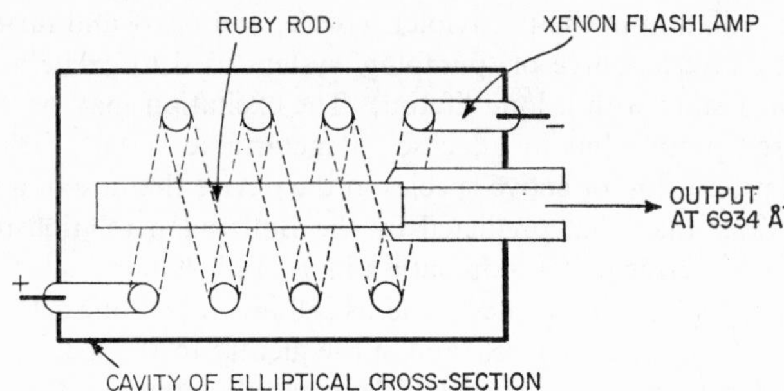

Figure A1.2 The ruby laser; a potential of several thousand volts is applied across the electrodes, and when the flash is required a very high potential is pulsed to an initiator electrode

The laser process develops very rapidly, so the population inverted state is destroyed almost immediately after the process commences. However, the flashlamp source has a relatively long emission and therefore the inversion is regenerated only to be exhausted by new laser action. If one reflector of the optical cavity transmits very weakly it is possible to 'see' the process occurring with a photocell and oscilloscope, i.e. the output is 'spiked' (see Figure A1.3). To sum up, the output of a pulsed laser of the ruby type consists of a series of very short pulses ($\sim 10^{-8}$ sec each) over a period of about 1 millisecond. The radiation is coherent (all the wavefronts are in-phase) and also parallel (since only beams passing repeatedly up and down the cavity are fully amplified). Also, to satisfy the wavefront criteria the radiation must be very monochromatic. Typical output characteristics for a small ruby laser are as follows: wavelength 6934 Å; ruby dimensions 3 in $\times \frac{1}{4}$ in; threshold pulse energy from xenon flashlamp $\sim$200 J; laser output 0·1 J; beam divergence 0·003 radian.

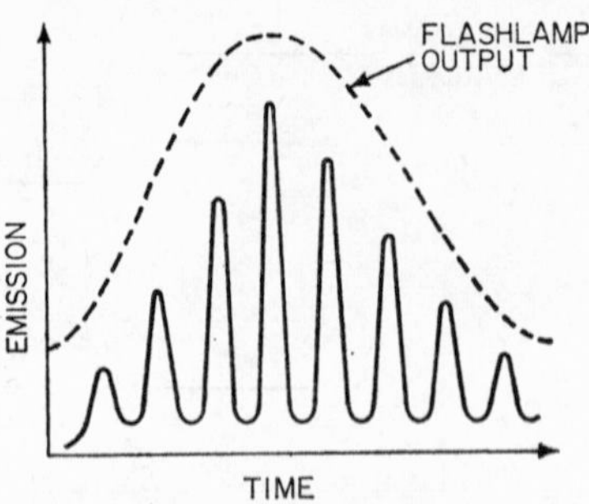

Figure A1.3 Spiking

Numerous other systems have been produced which are related to the ruby device, e.g. Nd^{3+} in yttrium aluminium garnet (1·046 μ output), Sm^{2+} in CaF_2, etc. (but none has ousted the ruby from its position of pre-eminence).

The Continuous Laser

The pulsed laser has been joined by a large number of devices which will operate continuously. With adequate design (especially efficient cooling) and use of very powerful sources such as the tungsten–iodine-in-quartz incandescent lamps, it is possible to operate some of the normal pulsed lasers in a continuous manner. The Nd^{3+} in yttrium aluminium garnet and ruby devices are typical in this respect, and outputs of the order of 1 watt have been obtained with them. By far the most important continuous lasers are those based on gaseous discharges. The classical device is the helium–neon laser which consists of a discharge tube containing the gases (He : Ne pressure ~1 : 7) inside a resonator cavity. The discharge tube is normally, but not always, closed with windows set at the Brewster angle (in which case radiation with electric vector in the plane of the normal to the window is very efficiently transmitted). As a result, the radiation produced by such a device is plane-polarized as well as being coherent. The cavity is generated normally by using either one plane and one convex or two convex mirrors. The energy level diagram for the device is given in Figure A1.4 and a typical laser is diagrammatically shown in Figure A1.5. Typical characteristics for a powerful helium–neon laser of value in Raman spectroscopy are: wavelength 6328 Å; linewidth ~0·1 cm^{-1}; power 30—150 mW ($\pm$0·5%, or less, fluctuation); divergence of beam $\ll 1°$; power consumption ~100 W; cooling by air. Lifetimes are now very extended; Spectra Physics have

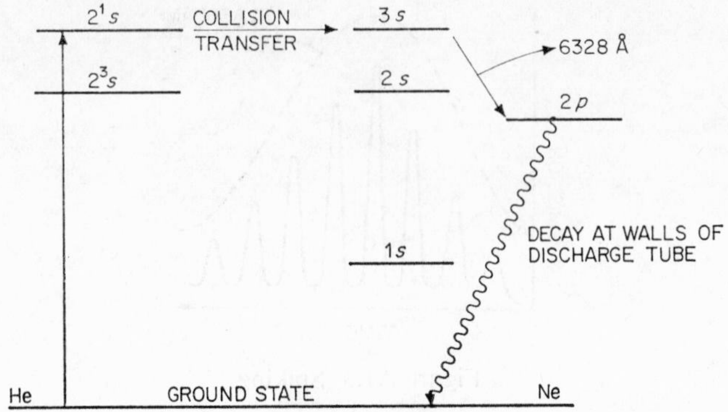

Figure A1.4 Energy levels (simplified) involved in the He–Ne laser; note that He atoms are in excess to favour collision of He (excited in 2^1s state) and Ne (ground state) atoms

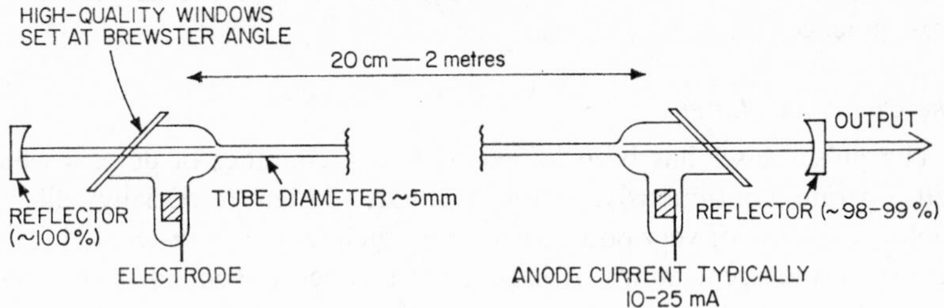

Figure A1.5 The helium–neon laser; a potential of several thousand volts d.c. and a cold-cathode system are frequently used

reported a tube life of >15,000 hr. In addition to this device, a range of much more powerful lasers have been developed which rely on the production of rare-gas cations. Probably the most widespread is the argon ion device but Kr^+ and other systems are available and are mentioned in Chapter 2. The layout of the Ar^+ laser is given in Figure A1.6. Typical characteristics are: wavelength of most powerful lines 4880 or 5145 Å; linewidth ~0·25 cm^{-1}; power 1—2 W per line (±2—10% fluctuation); divergence <1°; power *consumption* up to 10 kW normal; water cooling; life of tubes ~1000 hr. In order to generate large numbers of cations, an ultra-intense discharge is produced (0·3 cm diameter tubing often used to carry a current of 30 amp!) in the cavity. To confine laser action to one line, a prism is usually placed inside the cavity and arranged so that rotation thereof makes the cavity a resonator for only one wavelength.

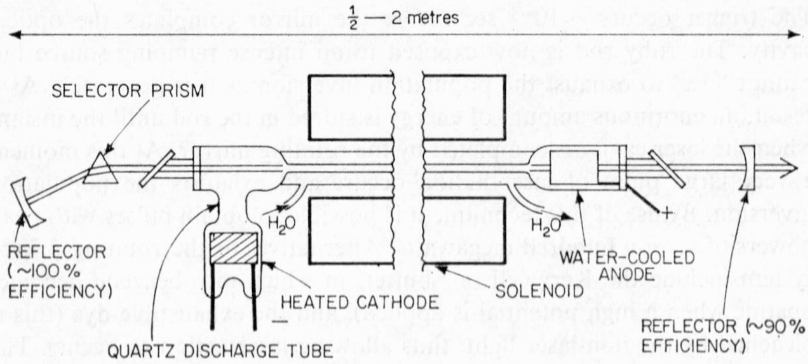

Figure A1.6 A d.c. powered argon ion laser; a small-bore tube is usually incorporated, to return the discharged argon from the cathode to the anode compartment

The Giant Pulse or Q-Spoiled Laser

This is a variant of the familiar ruby laser. The intention is to prevent the spiking phenomenon and to compress all the energy output into one enormous pulse. There are a number of ways of achieving this effect but probably the rotating mirror system is the easiest to explain. The laser consists of a ruby rod reflection-coated at only one end. The other end is anti-reflection coated. The optical cavity is completed by a plane mirror capable of rotation about a vertical axis. When suitably aligned with the rod and its reflecting end, the laser can operate. The rotating mirror is set in motion and is used to trigger the xenon flashlamp around the ruby rod.

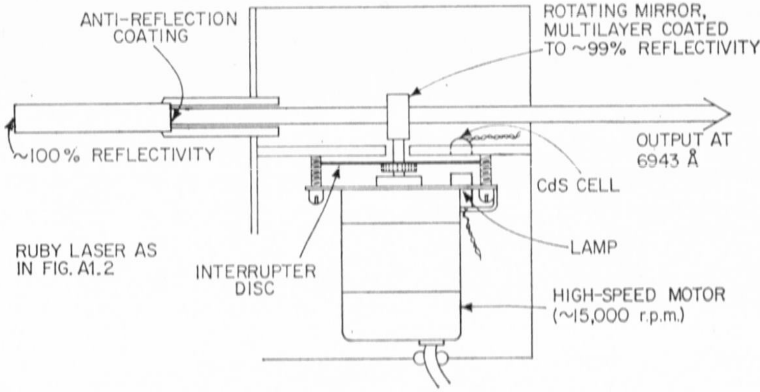

Figure A1.7 A design for a Q-spoiled ruby laser

The trigger occurs $\sim 10^{-3}$ sec *before* the mirror completes the optical cavity. The ruby rod is now exposed to an intense pumping source but cannot 'lase' to exhaust the population inversion as it is generated. As a result, an enormous amount of energy is stored in the rod until the instant when the laser cavity is completed by the rotating mirror. At this moment a very large pulse of laser action occurs and exhausts the population inversion. By use of this technique it is possible to obtain pulses with peak powers of several hundred megawatts. Alternatives to the rotating mirror system include the Kerr cell (a 'shutter' in which nitrobenzene becomes opaque when a high potential is applied), and the exhaustible dye (this is bleached by the non-laser light, thus allowing laser action to occur). The duration of the output pulse is very short ($\sim 10^{-8}$ sec).

Appendix 2

Results from group theory

A discussion of the activity of the external modes of crystals is included here since part of the tables must be interpreted using it. For convenience, the character tables and Raman tensors are presented in the conventional order for crystal point symmetries. Note that τ is used for (molecular) rotation, and α and β refer to polarizability and first hyperpolarizability respectively.

A2.1. Tensor and Vector Activity of Translations and Rotations

It is necessary to draw a clear distinction between the activity of rotations in 'free' species and in crystal sites, where the rotations generally become librational modes; similarly, translations are not in the normal sense optically active except in crystal sites. This arises because a free molecule carries its frame of reference with it, whereas in a site it is the external site axes which are important. The result is that the appearance of a rotation, together with vector or tensor components in a symmetry species, is a necessary but not a sufficient criterion for infrared or Raman activity of pure rotation in a free molecule. In addition, it is necessary for infrared (microwave) activity that the molecule should have a permanent dipole, which is re-oriented by the rotation in question. Thus, infrared activity is restricted to τ_x and τ_y of point groups C_n and C_{nv}, and to τ_x, τ_y, and τ_z of C_1 and C_s. For Raman activity of pure rotations, a molecule must present a varying aspect of the polarizability ellipsoid during the rotation. This is restricted to τ_x, τ_y, and τ_z of C_1, C_s, C_i, C_2, D_2, C_{2v}, C_{2h}, and D_{2h}, and to τ_x and τ_y of the remaining non-cubic point groups.

These more stringent restrictions may be relaxed in crystal sites, but relatively weak activity should result thereby. It is therefore interesting to determine the librational equivalents of the more stringent selection rules. Classically the appearance of infrared or Raman activity in an *oscillator* arises from a limiting non-zero value of $\delta\mu/\delta q_i$ or $\delta\alpha/\delta q_i$ as $\delta q_i \to 0$, where δq_i is measured *from the equilibrium position*. [This is simply a definition of the more usual formulation $(\partial\mu/\partial q_i)_0 \neq 0$ or $(\partial\alpha/\partial q_i)_0 \neq 0$.]

For a molecule with permanent dipole μ_z, $(\partial\mu_x/\partial\tau_y)_0 = (\partial\mu_y/\partial\tau_x)_0 \neq 0$.* For a molecule with a triaxial polarizability ellipsoid, in which the cartesian axes are the three principal axes of the ellipsoid (normally D_2, C_{2v}, and D_{2h}, but also C_1, C_s, C_i, C_2, and C_{2h} if the cartesians are so chosen), only $(\partial\alpha_{yz}/\partial\tau_x)_0$, $(\partial\alpha_{xz}/\partial\tau_y)_0$, and $(\partial\alpha_{xy}/\partial\tau_z)_0$ are non-zero. For a molecule with a polarizability 'ellipsoid of rotation', with z as the unique axis (remaining non-cubic point groups; note that listed alternative axes for these point groups relate only to the xy-plane) $(\partial\alpha_{xz}/\partial\tau_y)_0 = (\partial\alpha_{yz}/\partial\tau_x)_0 \neq 0$ only. In applying the oriented gas phase approximation to librational modes, we note therefore that only the above components should be used. This result is always compatible with, but may be more stringent than, the ordinary symmetry restrictions on such librations. Where only one axis (normally z) of a triaxial polarizability ellipsoid may be located with certainty (C_s, C_2, and C_{2h}) the components $(\partial\alpha_{xx}/\partial\tau_z)_0$ and $(\partial\alpha_{yy}/\partial\tau_z)_0$ may also be non-zero. Exceptionally also, $(\partial\alpha_{xy}/\partial\tau_z)_0 = 0$. Where the ellipsoid axes are entirely unknown (C_1, C_i) the general tensor shown for vibrations applies also to librations.

These considerations indicate that the rotation/tensor species sometimes given for point group S_8 (τ_x, τ_y in e_1; xz, yz in e_3) must be in error. This could also have been determined by reference to D_{4d}, of which S_8 is a subgroup.

Comparatively few molecules and ions have an appreciable permanent dipole, and, probably for this reason, it has been suggested that librational modes are generally more noticeable in the Raman than the infrared. In an 'oriented gas phase' type of approximation, *translational* external modes of molecular crystals appear in neither effect. However, infrared activity (often of great intensity) appears for ionic crystals.

A2.2 Character Tables and Raman Tensors

The orientation of cartesian axes is listed for all groups and axes where a fixed location is possible. Where parameters θ and ϕ in the listed Raman tensors refer to a rotation of axes, the axes shown in the character table refer to the $\theta = 0$ and $\phi = 0$ Raman tensors. This is also the conventional location of axes, except in the case of the trigonal point groups for which $y \| \sigma_v$ and/or $\perp C_2$ has also been in use.

The orientations of degenerate tensor components are indicated in a variety of ways: (*a*) where a vector is in correlation, the corresponding component is symmetric with respect to a plane or twofold axis parallel to

---

*For a non-symmetric top, these two terms are equal only for a rather unusual normalization of the rotational coordinate.

the vector; (*b*) where a rotation is in correlation, the component is symmetric to a plane, but antisymmetric to a two-fold rotation, parallel to the top axis; (*c*) by appeal to subgroups (cubic groups); (*d*) directly.

Where the axis rotations θ and ϕ are in use as additional non-zero parameters, the value zero may be used for one of them where they occur in the same tensor, or for both where they appear in separate tensors. However, it is not possible to predict the orientation of the degeneracy that this represents, and vector correlations become inapplicable. Vector to rotor correlations given for such groups are valid only under the oriented gas phase approximation.

In general, axis rotations are given only where vector/tensor correlations are affected, although some unaffected cases may appear under a rotatable tensor, for convenience of including them in their proper section. Where any tensor appears with a rotation, it may be assumed that other members of the same set, not explicitly stated with an angular parameter, do not suffer any change of correlation under the given rotation. No more general rotation should be assumed however. In one case (*e* of D_{2d}) the *vector* is stated with a rotation; thus $\begin{pmatrix} \sin 2\theta \\ \cos 2\theta \\ 0 \end{pmatrix}$ is the usual matrix notation for $x \sin 2\theta + y \cos 2\theta$. It should be understood that this is purely a matter of convenience.

A2.2.1 C_1 and C_i (1 and $\bar{1}$; triclinic point symmetries)

C_1	E
A	1

C_i	E	i		
A_g	1	1	$\tau_x,\ \tau_y,\ \tau_z$	α
A_u	1	-1	x, y, z	β

$$\begin{pmatrix} a & d & e \\ d & b & f \\ e & f & c \end{pmatrix}$$

	$\tau_x,\ \tau_y,\ \tau_z$
C_1	$a(x,y,z)$
C_i	a_g

A2.2.2 C_2, C_s, and C_{2h} (2, *m*, and 2/*m*; monoclinic point symmetries)

C_2	E	$C_2(z)$		
A	1	1	$z,\ \tau_z$	α,β
B	1	-1	$x,\ y,\ \tau_x,\ \tau_y$	α,β

C_s	E	$\sigma_h(xy)$		
A'	1	1	$x,\ y,\ \tau_z$	α,β
A''	1	−1	$z,\ \tau_x,\ \tau_y$	α,β

C_{2h}	E	$C_2(z)$	i	$\sigma_h(xy)$		
A_g	1	1	1	1	τ_z	α
B_g	1	−1	1	−1	$\tau_x,\ \tau_y$	α
A_u	1	1	−1	−1	z	β
B_u	1	−1	−1	1	$x,\ y$	β

$$\begin{pmatrix} a & d & \\ d & b & \\ & & c \end{pmatrix} \qquad \begin{pmatrix} & & e \\ & & f \\ e & f & \end{pmatrix}$$

	τ_z	$\tau_x,\ \tau_y$
C_2	$a(z)$	$b(x,y)$
C_s	$a'(x,y)$	$a''(z)$
C_{2h}	a_g	b_g

A2.2.3 D_2, C_{2v}, and D_{2h} (222, *mm*2, and *mmm*; orthorhombic point symmetries)

D_2	E	$C_2(z)$	$C_2(y)$	$C_2(x)$		
A	1	1	1	1		α,β
B_1	1	1	−1	−1	$z,\ \tau_z$	α,β
B_2	1	−1	1	−1	$y,\ \tau_y$	α,β
B_3	1	−1	−1	1	$x,\ \tau_x$	α,β

C_{2v}	E	$C_2(z)$	$\sigma_v(xz)$	$\sigma_v'(yz)$		
A_1	1	1	1	1	z	α,β
A_2	1	1	−1	−1	τ_z	α,β
B_1	1	−1	1	−1	$x,\ \tau_y$	α,β
B_2	1	−1	−1	1	$y,\ \tau_x$	α,β

D_{2h}	E	$C_2(z)$	$C_2(y)$	$C_2(x)$	i	$\sigma(xy)$	$\sigma(xz)$	$\sigma(yz)$		
A_g	1	1	1	1	1	1	1	1		α
B_{1g}	1	1	−1	−1	1	1	−1	−1	τ_z	α
B_{2g}	1	−1	1	−1	1	−1	1	−1	τ_y	α
B_{3g}	1	−1	−1	1	1	−1	−1	1	τ_x	α
A_u	1	1	1	1	−1	−1	−1	−1		β
B_{1u}	1	1	−1	−1	−1	−1	1	1	z	β
B_{2u}	1	−1	1	−1	−1	1	−1	1	y	β
B_{3u}	1	−1	−1	1	−1	1	1	−1	x	β

$$\begin{pmatrix} a & & \\ & b & \\ & & c \end{pmatrix} \quad \underset{\tau_z}{\begin{pmatrix} & d & \\ d & & \\ & & \end{pmatrix}} \quad \underset{\tau_y}{\begin{pmatrix} & & e \\ & & \\ e & & \end{pmatrix}} \quad \underset{\tau_x}{\begin{pmatrix} & & \\ & & f \\ & f & \end{pmatrix}}$$

D_2	a	$b_1(z)$	$b_2(y)$	$b_3(x)$
C_{2v}	$a_1(z)$	a_2	$b_1(x)$	$b_2(y)$
D_{2h}	a_g	b_{1g}	b_{2g}	b_{3g}

For orthorhombic symmetries, choice of symmetry elements parallel to all axes (D_2, D_{2h}) *or* x and y (C_{2v}) is arbitrary, and the b species notation follows from the choice.

A2.2.4 C_4, S_4, and C_{4h} (4, $\bar{4}$, and $4/m$; some tetragonal point symmetries)

C_4	E	$C_4(z)$	C_2	C_4^3		
A	1	1	1	1	z, τ_z	α,β
B	1	−1	1	−1		α,β
E	$\begin{Bmatrix}1 \\ 1\end{Bmatrix}$	$\begin{matrix}i \\ -i\end{matrix}$	$\begin{matrix}-1 \\ -1\end{matrix}$	$\begin{matrix}-i \\ i\end{matrix}$	(x,y) (τ_x, τ_y)	α,β

S_4	E	$S_4(z)$	C_2	S_4^3		
A	1	1	1	1	τ_z	α,β
B	1	−1	1	−1	z	α,β
E	$\begin{Bmatrix}1 \\ 1\end{Bmatrix}$	$\begin{matrix}i \\ -i\end{matrix}$	$\begin{matrix}-1 \\ -1\end{matrix}$	$\begin{matrix}-i \\ i\end{matrix}$	(x,y) (τ_x, τ_y)	α,β

C_{4h}	E	$C_4(z)$	C_2	C_4^3	i	S_4^3	$\sigma_h(xy)$	S_4		
A_g	1	1	1	1	1	1	1	1	τ_z	α
B_g	1	−1	1	−1	1	−1	1	−1		α
E_g	1	i	−1	$-i$	1	i	−1	$-i$	(τ_x, τ_y)	α
	1	$-i$	−1	i	1	$-i$	−1	i		
A_u	1	1	1	1	−1	−1	−1	−1	z	β
B_u	1	−1	1	−1	−1	1	−1	1		β
E_u	1	i	−1	$-i$	−1	$-i$	1	i	(x,y)	β
	1	$-i$	−1	i	−1	i	1	$-i$		

$$\begin{pmatrix} a & & \\ & a & \\ & & b \end{pmatrix} \quad \begin{pmatrix} c & d & \\ d & -c & \\ & & \end{pmatrix} \quad \begin{pmatrix} & & e\cos\phi \\ & & e\sin\phi \\ e\cos\phi & e\sin\phi & \end{pmatrix} \quad \begin{pmatrix} & & -e\sin\phi \\ & & e\cos\phi \\ -e\sin\phi & e\cos\phi & \end{pmatrix}$$

	τ_z		τ_y	τ_x
C_4	$a(z)$	b	$e(x)$	$e(y)$
S_4	a	$b(z)$	$e(y)$	$e(x)$
C_{4h}	a_g	b_g	e_g	e_g

(ϕ is an additional variable parameter)

For S_4, note that with no permanent dipole there is no o.g.p. correlation for vector even in librations.

A2.2.5 D_4, C_{4v}, D_{2d}, and D_{4h} (422, 4*mm*, $\bar{4}2m$, and 4/*mmn*; remaining tetragonal point symmetries)

D_4	E	$2C_4(z)$	$C_2(\equiv C_4^2)$	$2C_2'(x,y)$	$2C_2''$		
A_1	1	1	1	1	1		α
A_2	1	1	1	−1	−1	z, τ_z	β
B_1	1	−1	1	1	−1		α,β
B_2	1	−1	1	−1	1		α,β
E	2	0	−2	0	0	(x,y) (τ_x, τ_y)	α,β

C_{4v}	E	$2C_4(z)$	C_2	$2\sigma_v(xz,yz)$	$2\sigma_d$		
A_1	1	1	1	1	1	z	α,β
A_2	1	1	1	−1	−1	τ_z	
B_1	1	−1	1	1	−1		α,β
B_2	1	−1	1	−1	1		α,β
E	2	0	−2	0	0	(x,y) (τ_x, τ_y)	α,β

D_{2d}	E	$2S_4(z)$	$C_2(\equiv S_4^2)$	$2C_2'(x,y)$	$2\sigma_d$		
A_1	1	1	1	1	1		α,β
A_2	1	1	1	−1	−1	τ_z	β
B_1	1	−1	1	1	−1		α
B_2	1	−1	1	−1	1	z	α,β
E	2	0	−2	0	0	(x,y) (τ_x, τ_y)	α,β

D_{4h}	E	$2C_4(z)$	C_2	$2C_2'(x,y)$	$2C_2''$	i	$2S_4$	σ_h	$2\sigma_v(xz,yz)$	$2\sigma_d$		
A_{1g}	1	1	1	1	1	1	1	1	1	1		α
A_{2g}	1	1	1	−1	−1	1	1	1	−1	−1	τ_z	
B_{1g}	1	−1	1	1	−1	1	−1	1	1	−1		α
B_{2g}	1	−1	1	−1	1	1	−1	1	−1	1		α
E_g	2	0	−2	0	0	2	0	−2	0	0	(τ_x,τ_y)	α
A_{1u}	1	1	1	1	1	−1	−1	−1	−1	−1		
A_{2u}	1	1	1	−1	−1	−1	−1	−1	1	1	z	β
B_{1u}	1	−1	1	1	−1	−1	1	−1	−1	1		β
B_{2u}	1	−1	1	−1	1	−1	1	−1	1	−1		β
E_u	2	0	−2	0	0	−2	0	2	0	0	(x,y)	β

	$\begin{pmatrix} a & & \\ & a & \\ & & b \end{pmatrix}$	$\begin{pmatrix} c\cos 2\theta & c\sin 2\theta & \\ c\sin 2\theta & -c\cos 2\theta & \\ & & \end{pmatrix}$		$\begin{pmatrix} -d\sin 2\theta & d\cos 2\theta & \\ d\cos 2\theta & d\sin 2\theta & \\ & & \end{pmatrix}$	$\begin{pmatrix} & & e \\ & & \\ e & & \end{pmatrix}$ τ_y	$\begin{pmatrix} & & \\ & & e \\ & e & \end{pmatrix}$ τ_x
D_4	a_1	b_1	$x,y \parallel C'_2, \theta = 0$ $x,y \parallel C''_2, \theta = \pi/4$	b_2	$e(-y)$	$e(x)$
O	$a_1(b = a)$ $e(a = -\frac{1}{2}b = c/\sqrt{3})$	e		$f_2(d = e)$	f_2	f_2
C_{4v}	$a_1(z)$	b_1	$x,y \parallel \sigma_v, \theta = 0$ $x,y \parallel \sigma_d, \theta = \pi/4$	b_2	$e(x)$	$e(y)$
D_{2d}	a_1	b_1	$x,y \parallel C'_2, \theta = 0$ $x,y \parallel \sigma_d, \theta = \pi/4$	$b_2(z)$	$e \begin{pmatrix} \sin 2\theta \\ \cos 2\theta \\ 0 \end{pmatrix}$	$e \begin{pmatrix} \cos 2\theta \\ -\sin 2\theta \\ 0 \end{pmatrix}$
T_d	$a_1(b = a)$ $e(a = -\frac{1}{2}b = c/\sqrt{3})$	e		$f_2(z)\ (d = e)$	f_2 ,,	f_2 ,,
D_{4h}	a_{1g}	b_{1g}	$x,y \parallel C'_2, \sigma_v, \theta = 0$ $x,y \parallel C''_2, \sigma_d, \theta = \pi/4$	b_{2g}	e_g	e_g
O_h	$a_{1g}(b = a)$ $e_g(a = -\frac{1}{2}b = c/\sqrt{3})$	e_g		$f_{2g}(d = e)$	f_{2g}	f_{2g}

For those tetragonal groups (D_4, C_{4v}, D_{4h}) in which the choice $\theta = 0$ or $\pi/4$ is apparently quite arbitrary, it is strictly correct to change the species notation b_1/b_2 for $\theta = \pi/4$. We resolve this ambiguity by distinguishing strict point group ($\theta = 0$) axes from variable laboratory or observational axes.

A2.2.6 C_3, S_6, D_3, C_{3v}, and D_{3d} (3, $\bar{3}$, 32, $3m$, and $\bar{3}m$; trigonal point symmetries)

C_3	E	$C_3(z)$	C_3^2		$\varepsilon = \exp(2\pi i/3)$
A	1	1	1	z, τ_z	α,β
E	$\begin{Bmatrix} 1 \\ 1 \end{Bmatrix}$	$\begin{matrix} \varepsilon \\ \varepsilon^* \end{matrix}$	$\begin{matrix} \varepsilon^* \\ \varepsilon \end{matrix}$	(x,y) (τ_x, τ_y)	α,β

S_6	E	$C_3(z)$	C_3^2	i	S_6^5	S_6		$\varepsilon = \exp(2\pi i/3)$
A_g	1	1	1	1	1	1	τ_z	α
E_g	$\begin{Bmatrix} 1 \\ 1 \end{Bmatrix}$	$\begin{matrix} \varepsilon \\ \varepsilon^* \end{matrix}$	$\begin{matrix} \varepsilon^* \\ \varepsilon \end{matrix}$	$\begin{matrix} 1 \\ 1 \end{matrix}$	$\begin{matrix} \varepsilon \\ \varepsilon^* \end{matrix}$	$\begin{matrix} \varepsilon^* \\ \varepsilon \end{matrix}$	(τ_x, τ_y)	α
A_u	1	1	1	-1	-1	-1	z	β
E_u	$\begin{Bmatrix} 1 \\ 1 \end{Bmatrix}$	$\begin{matrix} \varepsilon \\ \varepsilon^* \end{matrix}$	$\begin{matrix} \varepsilon^* \\ \varepsilon \end{matrix}$	$\begin{matrix} -1 \\ -1 \end{matrix}$	$\begin{matrix} -\varepsilon \\ -\varepsilon^* \end{matrix}$	$\begin{matrix} -\varepsilon^* \\ -\varepsilon \end{matrix}$	(x,y)	β

D_3	E	$2C_3(z)$	$3C_2(y)$		
A_1	1	1	1		α,β
A_2	1	1	-1	z, τ_z	β
E	2	-1	0	(x,y) (τ_x, τ_y)	α,β

C_{3v}	E	$2C_3(z)$	$3\sigma_v(xz)$		
A_1	1	1	1	z	α,β
A_2	1	1	-1	τ_z	β
E	2	-1	0	(x,y) (τ_x, τ_y)	α,β

D_{3d}	E	$2C_3(z)$	$3C_2(y)$	i	$2S_6$	$3\sigma_d(xz)$		
A_{1g}	1	1	1	1	1	1		α
A_{2g}	1	1	-1	1	1	-1	τ_z	
E_g	2	-1	0	2	-1	0	(τ_x, τ_y)	α
A_{1u}	1	1	1	-1	-1	-1		β
A_{2u}	1	1	-1	-1	-1	1	z	β
E_u	2	-1	0	-2	1	0	(x,y)	β

Group	$\begin{pmatrix} a & \\ & b \end{pmatrix}$	$\begin{pmatrix} c \sin 3\theta & -c \cos 3\theta & d \sin \phi \\ d \cos \phi & d \sin \phi & \end{pmatrix}$ τ_y		$\begin{pmatrix} -c \cos 3\theta & -c \sin 3\theta & d \cos \phi \\ -d \sin \phi & d \cos \phi & \end{pmatrix}$ τ_x
C_3	$a(z)\ (\tau_z)$	$e(x)$	θ and ϕ variable	$e(y)$
T	$a(b = a)$ $f(a = -\frac{1}{2}b = d)\ (z)\ (\tau_z)$	$e(d = c\sqrt{2})$ (not τ) $f(c = -d\sqrt{2})\ (x)$	x in $zC_2, \theta = 0$ y in $zC_2, \theta = \pi/6$ $\phi = 0$	$e(d = c\sqrt{2})$ (not τ) $f(c = -d\sqrt{2})\ (y)$
$S_6 \equiv C_{3i}$	$a_g(\tau_z)$	e_g	θ and ϕ variable	e_g
T_h	$a_g(b = a)$ $f_g(a = -\frac{1}{2}b = d)\ (\tau_z)$	$e_g(d = c\sqrt{2})$ (not τ) $f_g(c = -d\sqrt{2})$	x in $zC_2, \theta = 0$ y in $zC_2, \theta = \pi/6$ $\phi = 0$	$e_g(d = c\sqrt{2})$ (not τ) $f_g(c = -d\sqrt{2})$
D_3	a_1	$e(-y)$	$y \parallel C_2, \theta = 0$ $x \parallel C_2, \theta = \pi/6$ $\phi = 0$	$e(x)$
O	$a_1(b = a)$ $f_2(a = -\frac{1}{2}b = d)$	$e(d = c\sqrt{2})$ $f_2(c = -d\sqrt{2})$ (not τ)		$e(d = c\sqrt{2})$ $f_2(c = -d\sqrt{2})$ (not τ)
C_{3v}	$a_1(z)$	$e(x)$	$x \parallel \sigma_v, \theta = 0$ $y \parallel \sigma_v, \theta = \pi/6$ $\phi = 0$	$e(y)$
T_d	$a_1(b = a)$ $f_2(a = -\frac{1}{2}b = d)\ (z)$	$e(d = c\sqrt{2})$ $f_2(c = -d\sqrt{2})$ (not τ)		$e(d = c\sqrt{2})$ $f_2(c = -d\sqrt{2})$ (not τ)
D_{3d}	a_{1g}	e_g	$y \parallel C_2, \perp \sigma_d$ $\theta = 0$ $x \parallel C_2, \perp \sigma_d$ $\theta = \pi/6$ $\phi = 0$	e_g
O_h	$a_{1g}(b = a)$ $f_{2g}(a = -\frac{1}{2}b = d)$	$e_g(d = c\sqrt{2})$ $_{2g}(c = -d\sqrt{2})$ (not τ)		$e_g(d = c\sqrt{2})$ $f_{2g}(c = -d\sqrt{2})$ (not τ)

A2.2.7 C_6, C_{3h}, C_{6h}, D_{3h} (6, $\bar{6}$, $6/m$, $\bar{6}m2$; some hexagonal point symmetries) C_5, C_{5h}, C_7, C_{7h}, S_8

C_6	E	$C_6(z)$	C_3	C_2	C_3^2	C_6^5		$\varepsilon = \exp(2\pi i/6)$
A	1	1	1	1	1	1	z, τ_z	α,β
B	1	-1	1	-1	1	-1		β
E_1	1	ε	$-\varepsilon^*$	-1	$-\varepsilon$	ε^*	(x,y) (τ_x, τ_y)	α,β
	1	ε^*	$-\varepsilon$	-1	$-\varepsilon^*$	ε		
E_2	1	$-\varepsilon^*$	$-\varepsilon$	1	$-\varepsilon^*$	$-\varepsilon$		α,β
	1	$-\varepsilon$	$-\varepsilon^*$	1	$-\varepsilon$	$-\varepsilon^*$		

C_{3h}	E	$C_3(z)$	C_3^2	$\sigma_h(xy)$	S_3	S_3^5		$\varepsilon = \exp(2\pi i/3)$
A'	1	1	1	1	1	1	τ_z	α,β
E'	1	ε	ε^*	1	ε	ε^*	(x,y)	α,β
	1	ε^*	ε	1	ε^*	ε		
A''	1	1	1	-1	-1	-1	z	β
E''	1	ε	ε^*	-1	$-\varepsilon$	$-\varepsilon^*$	(τ_x, τ_y)	α,β
	1	ε^*	ε	-1	$-\varepsilon^*$	$-\varepsilon$		

C_{6h}	E	$C_6(z)$	C_3	C_2	C_3^2	C_6^5	i	S_3^5	S_6^5	$\sigma_h(xy)$	S_6	S_3		$\varepsilon = \exp(2\pi i/6)$
A_g	1	1	1	1	1	1	1	1	1	1	1	1	τ_z	α
B_g	1	-1	1	-1	1	-1	1	-1	1	-1	1	-1		
E_{1g}	1	ε	$-\varepsilon^*$	-1	$-\varepsilon$	ε^*	1	ε	$-\varepsilon^*$	-1	$-\varepsilon$	ε^*	(τ_x, τ_y)	α
	1	ε^*	$-\varepsilon$	-1	$-\varepsilon^*$	ε	1	ε^*	$-\varepsilon$	-1	$-\varepsilon^*$	ε		
E_{2g}	1	$-\varepsilon^*$	$-\varepsilon$	1	$-\varepsilon^*$	$-\varepsilon$	1	$-\varepsilon^*$	$-\varepsilon$	1	$-\varepsilon^*$	$-\varepsilon$		α
	1	$-\varepsilon$	$-\varepsilon^*$	1	$-\varepsilon$	$-\varepsilon^*$	1	$-\varepsilon$	$-\varepsilon^*$	1	$-\varepsilon$	$-\varepsilon^*$		
A_u	1	1	1	1	1	1	-1	-1	-1	-1	-1	-1	z	β
B_u	1	-1	1	-1	1	-1	-1	1	-1	1	-1	1		β
E_{1u}	1	ε	$-\varepsilon^*$	-1	$-\varepsilon$	ε^*	-1	$-\varepsilon$	ε^*	1	ε	$-\varepsilon^*$	(x,y)	β
	1	ε^*	$-\varepsilon$	-1	$-\varepsilon^*$	ε	-1	$-\varepsilon^*$	ε	1	ε^*	$-\varepsilon$		
E_{2u}	1	$-\varepsilon^*$	$-\varepsilon$	1	$-\varepsilon^*$	$-\varepsilon$	-1	ε^*	ε	-1	ε^*	ε		β
	1	$-\varepsilon$	$-\varepsilon^*$	1	$-\varepsilon$	$-\varepsilon^*$	-1	ε	ε^*	-1	ε	ε^*		

D_{3h}	E	$2C_3(z)$	$3C_2(x)$	σ_h	$2S_3$	$3\sigma_v(xz)$		
A_1'	1	1	1	1	1	1		α,β
A_2'	1	1	-1	1	1	-1	τ_z	β
E'	2	-1	0	2	-1	0	(x,y)	α,β
A_1''	1	1	1	-1	-1	-1		
A_2''	1	1	-1	-1	-1	1	z	β
E''	2	-1	0	-2	1	0	(τ_x, τ_y)	α,β

C_5	E	$C_5(z)$	C_5^2	C_5^3	C_5^4		$\varepsilon = \exp(2\pi i/5)$
A	1	1	1	1	1	z, τ_z	α,β
E_1	$\left\{\begin{matrix}1\\1\end{matrix}\right.$	$\begin{matrix}\varepsilon\\\varepsilon^*\end{matrix}$	$\begin{matrix}\varepsilon^2\\\varepsilon^{2*}\end{matrix}$	$\begin{matrix}\varepsilon^{2*}\\\varepsilon^2\end{matrix}$	$\left.\begin{matrix}\varepsilon^*\\\varepsilon\end{matrix}\right\}$	$(x,y)\ (\tau_x, \tau_y)$	α,β
E_2	$\left\{\begin{matrix}1\\1\end{matrix}\right.$	$\begin{matrix}\varepsilon^2\\\varepsilon^{2*}\end{matrix}$	$\begin{matrix}\varepsilon^*\\\varepsilon\end{matrix}$	$\begin{matrix}\varepsilon\\\varepsilon^*\end{matrix}$	$\left.\begin{matrix}\varepsilon^{2*}\\\varepsilon^2\end{matrix}\right\}$		α,β

C_{5h}	E	$C_5(z)$	C_5^2	C_5^3	C_5^4	$\sigma_h(xy)$	S_5	S_5^7	S_5^3	S_5^9		$\varepsilon = \exp(2\pi i/5)$
A'	1	1	1	1	1	1	1	1	1	1	τ_z	α
E_1'	1	ε	ε^2	ε^{2*}	ε^*	1	ε	ε^2	ε^{2*}	ε^*	(x,y)	β
	1	ε^*	ε^{2*}	ε^2	ε	1	ε^*	ε^{2*}	ε^2	ε		
E_2'	1	ε^2	ε^*	ε	ε^{2*}	1	ε^2	ε^*	ε	ε^{2*}		α,β
	1	ε^{2*}	ε	ε^*	ε^2	1	ε^{2*}	ε	ε^*	ε^2		
A''	1	1	1	1	1	-1	-1	-1	-1	-1	z	β
E_1''	1	ε	ε^2	ε^{2*}	ε^*	-1	$-\varepsilon$	$-\varepsilon^2$	$-\varepsilon^{2*}$	$-\varepsilon^*$	(τ_x, τ_y)	α
	1	ε^*	ε^{2*}	ε^2	ε	-1	$-\varepsilon^*$	$-\varepsilon^{2*}$	$-\varepsilon^2$	$-\varepsilon$		
E_2''	1	ε^2	ε^*	ε	ε^{2*}	-1	$-\varepsilon^2$	$-\varepsilon^*$	$-\varepsilon$	$-\varepsilon^{2*}$		β
	1	ε^{2*}	ε	ε^*	ε^2	-1	$-\varepsilon^{2*}$	$-\varepsilon$	$-\varepsilon^*$	$-\varepsilon^2$		

C_7	E	$C_7(z)$	C_7^2	C_7^3	C_7^4	C_7^5	C_7^6		$\varepsilon = \exp(2\pi i/7)$
A	1	1	1	1	1	1	1	z, τ_z	α,β
E_1	1	ε	ε^2	ε^3	ε^{3*}	ε^{2*}	ε^*	(x,y)	α,β
	1	ε^*	ε^{2*}	ε^{3*}	ε^3	ε^2	ε	(τ_x, τ_y)	
E_2	1	ε^2	ε^{3*}	ε^*	ε	ε^3	ε^{2*}		α,β
	1	ε^{2*}	ε^3	ε	ε^*	ε^{3*}	ε^2		
E_3	1	ε^3	ε^*	ε^2	ε^{2*}	ε	ε^{3*}		β
	1	ε^{3*}	ε	ε^{2*}	ε^2	ε^*	ε^3		

$C_{7h} = C_7 \times \sigma_h$

S_8	E	$S_8(z)$	C_4	S_8^3	C_2	S_8^5	C_4^3	S_8^7		$\varepsilon = \exp(2\pi i/8)$
A	1	1	1	1	1	1	1	1	τ_z	α
B	1	-1	1	-1	1	-1	1	-1	z	β
E_1	1	ε	i	$-\varepsilon^*$	-1	$-\varepsilon$	$-i$	ε^*	(x,y)	β
	1	ε^*	$-i$	$-\varepsilon$	-1	$-\varepsilon^*$	i	ε		
E_2	1	i	-1	$-i$	1	i	-1	$-i$		α,β
	1	$-i$	-1	i	1	$-i$	-1	i		
E_3	1	$-\varepsilon^*$	$-i$	ε	-1	ε^*	i	$-\varepsilon$	(τ_x, τ_y)	α,β
	1	$-\varepsilon$	i	ε^*	-1	ε	$-i$	$-\varepsilon^*$		

$$\begin{pmatrix} a & & \\ & a & \\ & & b \end{pmatrix} \begin{pmatrix} c\cos\theta & c\sin\theta & \\ c\sin\theta & -c\cos\theta & \\ & & \end{pmatrix} \begin{pmatrix} c\sin\theta & -c\cos\theta & \\ -c\cos\theta & -c\sin\theta & \\ & & \end{pmatrix} \begin{pmatrix} & & d\cos\phi \\ & & d\sin\phi \\ d\cos\phi & d\sin\phi & \end{pmatrix} \begin{pmatrix} & & -d\sin\phi \\ & & d\cos\phi \\ -d\sin\phi & d\cos\phi & \end{pmatrix}$$

	τ_z				τ_y		τ_x
C_6	$a(z)$	e_2	θ variable	e_2	$e_1(x)$	ϕ variable	$e_1(y)$
C_{3h}	a'	$e'(x)$		$e'(y)$	e''		e''
C_{6h}	a_g	e_{2g}		e_{2g}	e_{1g}		e_{1g}
D_{3h}	a_1' (not τ)	$e'(x)$	$\theta = 3\rho$; $x \parallel C_2$ & $\sigma_v, \rho = 0$; $y \parallel C_2$ & $\sigma_v, \rho = \pi/6$	$e'(y)$	e''	$\phi = 0$	e''
C_5, C_7	$a(z)$	e_2	θ variable	e_2	$e_1(x)$	ϕ variable	$e_1(y)$
C_{5h}, C_{7h}	a'	e_2'		e_2'	e_1''		e_1''
S_8	a	e_2		e_2	e_3		e_3

*A*2.2.8 D_6, C_{6v}, D_{6h} (622, 6*mm*, and 6/*mmm*; remaining hexagonal point symmetries)
D_5, C_{5v}, D_{5h}, D_{5d}
D_7, C_{7v}, D_{7h}, D_{7d}
D_{4d}, D_{6d}, $C_{\infty v}$, $D_{\infty h}$

D_6	E	$2C_6(z)$	$2C_3$	C_2	$3C_2'(y)$	$3C_2''(x)$		
A_1	1	1	1	1	1	1		α
A_2	1	1	1	1	−1	−1	z, τ_z	β
B_1	1	−1	1	−1	1	−1		β
B_2	1	−1	1	−1	−1	1		β
E_1	2	1	−1	−2	0	0	(x,y) (τ_x, τ_y)	α,β
E_2	2	−1	−1	2	0	0		α,β

C_{6v}	E	$2C_6(z)$	$2C_3$	C_2	$3\sigma_v(xz)$	$3\sigma_d(yz)$		
A_1	1	1	1	1	1	1	z	α,β
A_2	1	1	1	1	−1	−1	τ_z	
B_1	1	−1	1	−1	1	−1		β
B_2	1	−1	1	−1	−1	1		β
E_1	2	1	−1	−2	0	0	(x,y) (τ_x, τ_y)	α,β
E_2	2	−1	−1	2	0	0		α,β

D_{6h}	E	$2C_6(z)$	$2C_3$	C_2	$3C'_2(y)$	$3C''_2(x)$	i	$2S_3$	$2S_6$	σ_h	$3\sigma_d(xz)$	$3\sigma_v(yz)$		
A_{1g}	1	1	1	1	1	1	1	1	1	1	1	1		α
A_{2g}	1	1	1	1	−1	−1	1	1	1	1	−1	−1	τ_z	
B_{1g}	1	−1	1	−1	1	−1	1	−1	1	−1	1	−1		
B_{2g}	1	−1	1	−1	−1	1	1	−1	1	−1	−1	1		
E_{1g}	2	1	−1	−2	0	0	2	1	−1	−2	0	0	(τ_x, τ_y)	α
E_{2g}	2	−1	−1	2	0	0	2	−1	−1	2	0	0		α
A_{1u}	1	1	1	1	1	1	−1	−1	−1	−1	−1	−1		
A_{2u}	1	1	1	1	−1	−1	−1	−1	−1	−1	1	1	z	β
B_{1u}	1	−1	1	−1	1	−1	−1	1	−1	1	−1	1		β
B_{2u}	1	−1	1	−1	−1	1	−1	1	−1	1	1	−1		β
E_{1u}	2	1	−1	−2	0	0	−2	−1	1	2	0	0	(x,y)	β
E_{2u}	2	−1	−1	2	0	0	−2	1	1	−2	0	0		β

D_5	E	$2C_5(z)$	$2C_5^2$	$5C_2(x \text{ or } y)$		
A_1	1	1	1	1		α
A_2	1	1	1	−1	z, τ_z	β
E_1	2	2 cos 72°	2 cos 144°	0	(x,y) (τ_x, τ_y)	α,β
E_2	2	2 cos 144°	2 cos 72°	0		α,β

$2 \cos 72° = \frac{1}{2}(\sqrt{5}-1)$; $2 \cos 144° = -\frac{1}{2}(\sqrt{5}+1)$

C_{5v}	E	$2C_5(z)$	$2C_5^2$	$5\sigma_v(xz \text{ or } yz)$		
A_1	1	1	1	1	z	α,β
A_2	1	1	1	−1	τ_z	
E_1	2	2 cos 72°	2 cos 144°	0	(x,y) (τ_x, τ_y)	α,β
E_2	2	2 cos 144°	2 cos 72°	0		α,β

D_{5h}	E	$2C_5(z)$	$2C_5^2$	$5C_2$	σ_h	$2S_5$	$2S_5^3$	$5\sigma_v(xz \text{ or } yz)$		
A_1'	1	1	1	1	1	1	1	1		α
A_2'	1	1	1	−1	1	1	1	−1	τ_z	
E_1'	2	2 cos 72°	2 cos 144°	0	2	2 cos 72°	2 cos 144°	0	(x,y)	β
E_2'	2	2 cos 144°	2 cos 72°	0	2	2 cos 144°	2 cos 72°	0		α,β
A_1''	1	1	1	1	−1	−1	−1	−1		
A_2''	1	1	1	−1	−1	−1	−1	1	z	β
E_1''	2	2 cos 72°	2 cos 144°	0	−2	−2 cos 72°	−2 cos 144°	0	(τ_x, τ_y)	α
E_2''	2	2 cos 144°	2 cos 72°	0	−2	−2 cos 144°	−2 cos 72°	0		β

D_{5d}	E	$2C_5(z)$	$2C_5^2$	$5C_2$(x or y)	i	$2S_{10}^3$	$2S_{10}$	$5\sigma_d$(y or x)		
A_{1g}	1	1	1	1	1	1	1	1		α
A_{2g}	1	1	1	−1	1	1	1	−1	τ_z	
E_{1g}	2	2 cos 72°	2 cos 144°	0	2	2 cos 72°	2 cos 144°	0	(τ_x, τ_y)	α
E_{2g}	2	2 cos 144°	2 cos 72°	0	2	2 cos 144°	2 cos 72°	0		α
A_{1u}	1	1	1	1	−1	−1	−1	−1		
A_{2u}	1	1	1	−1	−1	−1	−1	1	z	β
E_{1u}	2	2 cos 72°	2 cos 144°	0	−2	−2 cos 72°	−2 cos 144°	0	(x,y)	β
E_{2u}	2	2 cos 144°	2 cos 72°	0	−2	−2 cos 144°	−2 cos 72°	0		β

D_{4d}	E	$2S_8(z)$	$2C_4$	$2S_8^3$	C_2	$4C_2'(x,y)$	$4\sigma_d$		
A_1	1	1	1	1	1	1	1		α
A_2	1	1	1	1	1	−1	−1	τ_z	
B_1	1	−1	1	−1	1	1	−1		
B_2	1	−1	1	−1	1	−1	1	z	β
E_1	2	$\sqrt{2}$	0	$-\sqrt{2}$	−2	0	0	(x,y)	β
E_2	2	0	−2	0	2	0	0		α,β
E_3	2	$-\sqrt{2}$	0	$\sqrt{2}$	−2	0	0	(τ_x, τ_y)	a,β

D_{6d}	E	$2S_{12}(z)$	$2C_6$	$2S_4$	$2C_3$	$2S_{12}^5$	C_2	$6C_2'(y)$	$6\sigma_d(xz)$		
A_1	1	1	1	1	1	1	1	1	1		α
A_2	1	1	1	1	1	1	1	-1	-1	τ_z	
B_1	1	-1	1	-1	1	-1	1	1	-1		
B_2	1	-1	1	-1	1	-1	1	-1	1	z	β
E_1	2	$\sqrt{3}$	1	0	-1	$-\sqrt{3}$	-2	0	0	(x,y)	β
E_2	2	1	-1	-2	-1	1	2	0	0		α
E_3	2	0	-2	0	2	0	-2	0	0		β
E_4	2	-1	-1	2	-1	-1	2	0	0		β
E_5	2	$-\sqrt{3}$	1	0	-1	$\sqrt{3}$	-2	0	0	(τ_x, τ_y)	α

$C_{\infty v}$	E	$2C_\infty^\Phi(z)$	...	$\infty\sigma_v$		
$A_1 = \Sigma^+$	1	1	...	1	z	α,β
$A_2 = \Sigma^-$	1	1	...	-1	τ_z	
$E_1 = \Pi$	2	$2\cos\Phi$	...	0	(x,y) (τ_x, τ_y)	α,β
$E_2 = \Delta$	2	$2\cos 2\Phi$	...	0		α,β
$E_3 = \Phi$	2	$2\cos 3\Phi$	...	0		β
...	...	...	...	...		

$D_{\infty h}$	E	$2C_\infty^\Phi(z)$	$\dots$	$\infty\sigma_v$	i	$2S_\infty^\Phi$	$\dots$	∞C_2		
Σ_g^+	1	1	$\dots$	1	1	1	$\dots$	1		α
Σ_g^-	1	1	$\dots$	-1	1	1	$\dots$	-1	τ_z	
Π_g	2	$2\cos\Phi$	$\dots$	0	2	$-2\cos\Phi$	$\dots$	0	(τ_x, τ_y)	α
Δ_g	2	$2\cos 2\Phi$	$\dots$	0	2	$2\cos 2\Phi$	$\dots$	0		α
$\dots$	$\dots$	$\dots$	$\dots$	$\dots$	$\dots$	$\dots$	$\dots$	$\dots$		
Σ_u^+	1	1	$\dots$	1	-1	-1	$\dots$	-1	z	β
Σ_u^-	1	1	$\dots$	-1	-1	-1	$\dots$	1		
Π_u	2	$2\cos\Phi$	$\dots$	0	-2	$2\cos\Phi$	$\dots$	0	(x,y)	β
Δ_u	2	$2\cos 2\Phi$	$\dots$	0	-2	$-2\cos 2\Phi$	$\dots$	0		β
Φ_u	2	$2\cos 3\Phi$	$\dots$	0	-2	$2\cos 3\Phi$	$\dots$	0		β
$\dots$	$\dots$	$\dots$	$\dots$	$\dots$	$\dots$	$\dots$	$\dots$	$\dots$		

For D_7 see C_5/D_5
For C_{7v} see C_5/C_{5v}
$D_{7h} = D_7 \times \sigma_h$
$D_{7d} = D_7 \times i$

	$\begin{pmatrix} a & & \\ & a & \\ & & b \end{pmatrix}$	$\begin{pmatrix} c & & \\ & -c & \\ & & \end{pmatrix}$		$\begin{pmatrix} & c & \\ c & & \\ & & \end{pmatrix}$		$\begin{pmatrix} & & d \\ & & \\ d & & \end{pmatrix}$ τ_y	$\begin{pmatrix} & & \\ & & d \\ & d & \end{pmatrix}$ τ_x
D_6	a_1	e_2	Symmetric with respect to any element parallel to x or y	e_2	Antisymmetric to elements parallel to x or y	$e_1(-y)$	$e_1(x)$
C_{6v}	$a_1(z)$	e_2		e_2		$e_1(x)$	$e_1(y)$
D_{6h}	a_{1g}	e_{2g}		e_{2g}		e_{1g}	e_{1g}
D_5, D_7	a_1	e_2		e_2		$e_1(-y)$	$e_1(x)$
C_{5v}, C_{7v}	$a_1(z)$	e_2		e_2		$e_1(x)$	$e_1(y)$
D_{5h}, D_{7h}	a_1'	e_2'		e_2'		e_1''	e_1''
D_{5d}, D_{7d}	a_{1g}	e_{2g}		e_{2g}		e_{1g}	e_{1g}
D_{4d}, S_8	a_1*	e_2		e_2		e_3	e_3
D_{6d}, S_{12}	a_1*	e_2		e_2		e_5	e_5
$C_{\infty v}$	$\Sigma^+(z)$	Δ		Δ		$\Pi(x)$	$\Pi(y)$
$D_{\infty h}$	Σ_g^+	Δ_g		Δ_g		Π_g	Π_g

*Drop subscript for S_n groups (species a only) giving alternative arbitrary orientations to A2.2.7.

A2.2.9 T, T_h, T_d, O, O_h (23, $m3$, $\bar{4}3m$, 432, and $m3m$; cubic point symmetries) I, I_h

T	E	$4C_3$	$4C_3^2$	$3C_2(x,y,z)$		$\varepsilon = \exp(2\pi i/3)$
A	1	1	1	1		α,β
E	1	ε	ε^*	1		α
	1	ε^*	ε	1		
F	3	0	0	-1	(x,y,z)	α,β
					(τ_x, τ_y, τ_z)	

T_h	E	$4C_3$	$4C_3^2$	$3C_2(x,y,z)$	i	$4S_6^5$	$4S_6$	$3\sigma_h(xy,xz,yz)$		$\varepsilon = \exp(2\pi i/3)$
A_g	1	1	1	1	1	1	1	1		α
A_u	1	1	1	1	-1	-1	-1	-1		β
E_g	1	ε	ε^*	1	1	ε	ε^*	1		α
	1	ε^*	ε	1	1	ε^*	ε	1		
E_u	1	ε	ε^*	1	-1	$-\varepsilon$	$-\varepsilon^*$	-1		
	1	ε^*	ε	1	-1	$-\varepsilon^*$	$-\varepsilon$	-1		
F_g	3	0	0	-1	3	0	0	-1	(τ_x, τ_y, τ_z)	α
F_u	3	0	0	-1	-3	0	0	1	(x,y,z)	β

T_d	E	$8C_3$	$3C_2(x,y,z)$	$6S_4$	$6\sigma_d$		
A_1	1	1	1	1	1		α,β
A_2	1	1	1	-1	-1		
E	2	-1	2	0	0		α
F_1	3	0	-1	1	-1	(τ_x, τ_y, τ_z)	β
F_2	3	0	-1	-1	1	(x,y,z)	α,β

O	E	$8C_3$	$6C_2'$	$6C_4$	$3C_2(=C_4^2)\ (x,y,z)$		
A_1	1	1	1	1	1		α
A_2	1	1	-1	-1	1		β
E	2	-1	0	0	2		α
F_1	3	0	-1	1	-1	(x,y,z) (τ_x, τ_y, τ_z)	β
F_2	3	0	1	-1	-1		α,β

O_h	E	$8C_3$	$6C'_2$	$6C_4$	$3C_2(=C_4^2)$ (x,y,z)	i	$6S_4$	$8S_6$	$3\sigma_h(xy,xz,yz)$	$6\sigma_d$		
A_{1g}	1	1	1	1	1	1	1	1	1	1		α
A_{2g}	1	1	−1	−1	1	1	−1	1	1	−1		
E_g	2	−1	0	0	2	2	0	−1	2	0		α
F_{1g}	3	0	−1	1	−1	3	1	0	−1	−1	(τ_x, τ_y, τ_z)	
F_{2g}	3	0	1	−1	−1	3	−1	0	−1	1		α
A_{1u}	1	1	1	1	1	−1	−1	−1	−1	−1		
A_{2u}	1	1	−1	−1	1	−1	1	−1	−1	1		β
E_u	2	−1	0	0	2	−2	0	1	−2	0		
F_{1u}	3	0	−1	1	−1	−3	−1	0	1	1	(x,y,z)	β
F_{2u}	3	0	1	−1	−1	−3	1	0	1	−1		β

$I_h = I x i$; for I drop g and u in I_h

I_h	E	$12C_5$	$12C_5^2$	$20C_3$	$15C_2$ (x,y,z)	i	$12S_{10}$	$12S_{10}^3$	$20S_6$	$15\sigma_h$ (xy,xz,yz)		
A_g	1	1	1	1	1	1	1	1	1	1		α
F_{1g}	3	$\frac{1}{2}(1+\sqrt{5})$	$\frac{1}{2}(1-\sqrt{5})$	0	−1	3	$\frac{1}{2}(1-\sqrt{5})$	$\frac{1}{2}(1+\sqrt{5})$	0	−1	(τ_x, τ_y, τ_z)	
F_{2g}	3	$\frac{1}{2}(1-\sqrt{5})$	$\frac{1}{2}(1+\sqrt{5})$	0	−1	3	$\frac{1}{2}(1+\sqrt{5})$	$\frac{1}{2}(1-\sqrt{5})$	0	−1		
G_g	4	−1	−1	1	0	4	−1	−1	1	0		
H_g	5	0	0	−1	1	5	0	0	−1	1		α
A_u	1	1	1	1	1	−1	−1	−1	−1	−1		(β not yet
F_{1u}	3	$\frac{1}{2}(1+\sqrt{5})$	$\frac{1}{2}(1-\sqrt{5})$	0	−1	−3	$-\frac{1}{2}(1-\sqrt{5})$	$-\frac{1}{2}(1+\sqrt{5})$	0	1	(x,y,z)	deter-
F_{2u}	3	$\frac{1}{2}(1-\sqrt{5})$	$\frac{1}{2}(1+\sqrt{5})$	0	−1	−3	$-\frac{1}{2}(1+\sqrt{5})$	$-\frac{1}{2}(1-\sqrt{5})$	0	1		mined)
G_u	4	−1	−1	1	0	−4	1	1	−1	0		
H_u	5	0	0	−1	1	−5	0	0	1	−1		

	$\begin{pmatrix} a & & \\ & a & \\ & & a \end{pmatrix}$	$\begin{pmatrix} b & & \\ & b & \\ & & -2b \end{pmatrix}$	$\begin{pmatrix} b\sqrt{3} & & \\ & -b\sqrt{3} & \\ & & \end{pmatrix}$	$\begin{pmatrix} & & \\ & & c \\ & c & \end{pmatrix}$	$\begin{pmatrix} & & c \\ & & \\ c & & \end{pmatrix}$	$\begin{pmatrix} & c & \\ c & & \\ & & \end{pmatrix}$
T	a	e	e	$f(x)\ (\tau_x)$	$f(y)\ (\tau_y)$	$f(z)\ (\tau_z)$
T_h	a_g	e_g	e_g	$f_g(\tau_x)$	$f_g(\tau_y)$	$f_g(\tau_z)$
T_d	a_1	e	e	$f_2(x)$	$f_2(y)$	$f_2(z)$
O	a_1	e	e	f_2	f_2	f_2
O_h	a_{1g}	e_g	e_g	f_{2g}	f_{2g}	f_{2g}
I	a	h	h	h	h	h } $c = b\sqrt{3}$
I_h	a_g	h_g	h_g	h_g	h_g	h_g }

Orientation information concerning the cubic Raman tensors is provided by reference to appropriate subgroups. The above are correct for $3C_2 = x,y,z$. Rotations about z (where intermediate xy-plane elements exist), and component orientation for $3C_2$ cartesians, are obtained by reference to the subgroups D_2, D_{2h}, D_{2d}, D_4, *and* D_{4h} respectively. Trigonal/hexagonal axes ($z \| C_3$) appear under subgroups C_3, S_6, C_{3v}, D_3, and D_{3d} respectively.

A2.3 Correlation Tables

Tables for the correlation of group and subgroup symmetry species. Correlations not directly given here may be obtained in two stages. The order is that of the point group character tables. Asterisks indicate separable degeneracy.

(i) C_{2h}, D_2, C_{2v}, D_{2h}, C_4, S_4, *and* C_{4h}.

C_{2h}	C_2	C_s	C_i
A_g	A	A'	A_g
B_g	B	A''	A_g
A_u	A	A''	A_u
B_u	B	A'	A_u

D_2	$C_2(z)$	$C_2(y)$	$C_2(x)$
A	A	A	A
B_1	A	B	B
B_2	B	A	B
B_3	B	B	A

C_{2v}	C_2	$\sigma(zx)$ C_s	$\sigma(yz)$ C_s
A_1	A	A'	A'
A_2	A	A''	A''
B_1	B	A'	A''
B_2	B	A''	A'

D_{2h}	D_2	$C_2(z)$† C_{2v}	$C_2(y)$† C_{2v}	$C_2(x)$† C_{2v}	$C_2(z)$ C_{2h}	$C_2(y)$ C_{2h}	$C_2(x)$ C_{2h}	$C_2(z)$ C_2	$C_2(y)$ C_2	$C_2(x)$ C_2	$\sigma(xy)$ C_s	$\sigma(zx)$ C_s	$\sigma(yz)$ C_s
A_g	A	A_1	A_1	A_1	A_g	A_g	A_g	A	A	A	A'	A'	A'
B_{1g}	B_1	A_2	B_2	B_1	A_g	B_g	B_g	A	B	B	A'	A''	A''
B_{2g}	B_2	B_1	A_2	B_2	B_g	A_g	B_g	B	A	B	A''	A'	A''
B_{3g}	B_3	B_2	B_1	A_2	B_g	B_g	A_g	B	B	A	A''	A''	A'
A_u	A	A_2	A_2	A_2	A_u	A_u	A_u	A	A	A	A''	A''	A''
B_{1u}	B_1	A_1	B_1	B_2	A_u	B_u	B_u	A	B	B	A''	A'	A'
B_{2u}	B_2	B_2	A_1	B_1	B_u	A_u	B_u	B	A	B	A'	A''	A'
B_{3u}	B_3	B_1	B_2	A_1	B_u	B_u	A_u	B	B	A	A'	A'	A''

† B species subscripts based on $\sigma(zx)$, $\sigma(yz)$, and $\sigma(xy)$ respectively. Interchange for other choice.

C_4	C_2
A	A
B	A
E^*	$2B$

S_4	C_2
A	A
B	A
E^*	$2B$

C_{4h}	C_4	S_4	C_{2h}	C_2	C_s	C_i
A_g	A	A	A_g	A	A'	A_g
B_g	B	B	A_g	A	A'	A_g
E_g^*	E^*	E^*	$2B_g$	$2B$	$2A''$	$2A_g$
A_u	A	B	A_u	A	A''	A_u
B_u	B	A	A_u	A	A''	A_u
E_u^*	E^*	E^*	$2B_u$	$2B$	$2A'$	$2A_u$

Correlation Tables (contd.): D_4, C_{4v}, D_{2d}, and D_{4h}.

D_4	C_4	C_2	C_2' C_2	C_2'' C_2
A_1	A	A	A	A
A_2	A	A	B	B
B_1	B	A	A	B
B_2	B	A	B	A
E	E^*	$2B$	$A+B$	$A+B$

C_{4v}	C_4	σ_v C_{2v}	σ_d C_{2v}	C_2	σ_v C_s	σ_d C_s
A_1	A	A_1	A_1	A	A'	A'
A_2	A	A_2	A_2	A	A''	A''
B_1	B	A_1	A_2	A	A'	A''
B_2	B	A_2	A_1	A	A''	A'
E	E^*	B_1+B_2	B_1+B_2	$2B$	$A'+A''$	$A'+A''$

D_{2d}	S_4	$C_2 \rightarrow C_2(z)$ D_2	C_{2v}	C_2 C_2	C_2' C_2	C_s
A_1	A	A	A_1	A	A	A'
A_2	A	B_1	A_2	A	B	A''
B_1	B	A	A_2	A	A	A''
B_2	B	B_1	A_1	A	B	A'
E	E^*	B_2+B_3	B_1+B_2	$2B$	$A+B$	$A'+A''$

D_{4h}	D_4	$C_2' \to C_2'$ D_{2d}	$C_2'' \to C_2'$ D_{2d}	C_{4v}	C_{4h}	C_2' D_{2h}	C_2'' D_{2h}	C_4	S_4	C_2' D_2	C_2'' D_2	C_2,σ_v C_{2v}	C_2,σ_d C_{2v}
A_{1g}	A_1	A_1	A_1	A_1	A_g	A_g	A_g	A	A	A	A	A_1	A_1
A_{2g}	A_2	A_2	A_2	A_2	A_g	B_{1g}	B_{1g}	A	A	B_1	B_1	A_2	A_2
B_{1g}	B_1	B_1	B_2	B_1	B_g	A_g	B_{1g}	B	B	A	B_1	A_1	A_2
B_{2g}	B_2	B_2	B_1	B_2	B_g	B_{1g}	A_g	B	B	B_1	A	A_2	A_1
E_g	E	E	E	E	E_g^*	$B_{2g}+B_{3g}$	$B_{2g}+B_{3g}$	E^*	E^*	B_2+B_3	B_2+B_3	B_1+B_2	B_1+B_2
A_{1u}	A_1	B_1	B_1	A_2	A_u	A_u	A_u	A	B	A	A	A_2	A_2
A_{2u}	A_2	B_2	B_2	A_1	A_u	B_{1u}	B_{1u}	A	B	B_1	B_1	A_1	A_1
B_{1u}	B_1	A_1	A_2	B_2	B_u	A_u	B_{1u}	B	A	A	B_1	A_2	A_1
B_{2u}	B_2	A_2	A_1	B_1	B_u	B_{1u}	A_u	B	A	B_1	A	A_1	A_2
E_u	E	E	E	E	E_u^*	$B_{2u}+B_{3u}$	$B_{2u}+B_{3u}$	E^*	E^*	B_2+B_3	B_2+B_3	B_1+B_2	B_1+B_2

D_{4h}	C_2'† C_{2v}	C_2''† C_{2v}	C_2 C_{2h}	C_2' C_{2h}	C_2'' C_{2h}	C_2 C_2	C_2' C_2	C_2'' C_2	σ_h C_s	σ_v C_s	σ_d C_s	C_i
A_{1g}	A_1	A_1	A_g	A_g	A_g	A	A	A	A'	A'	A'	A_g
A_{2g}	B_1	B_1	A_g	B_g	B_g	A	B	B	A'	A''	A''	A_g
B_{1g}	A_1	B_1	A_g	A_g	B_g	A	A	B	A'	A'	A''	A_g
B_{2g}	B_1	A_1	A_g	B_g	A_g	A	B	A	A'	A''	A'	A_g
E_g	A_2+B_2	A_2+B_2	$2B_g$	A_g+B_g	A_g+B_g	$2B$	$A+B$	$A+B$	$2A''$	$A'+A''$	$A'+A''$	$2A_g$
A_{1u}	A_2	A_2	A_u	A_u	A_u	A	A	A	A''	A''	A''	A_u
A_{2u}	B_2	B_2	A_u	B_u	B_u	A	B	B	A''	A'	A'	A_u
B_{1u}	A_2	B_2	A_u	A_u	B_u	A	A	B	A''	A''	A'	A_u
B_{2u}	B_2	A_2	A_u	B_u	A_u	A	B	A	A''	A'	A''	A_u
E_u	A_1+B_1	A_1+B_1	$2B_u$	A_u+B_u	A_u+B_u	$2B$	$A+B$	$A+B$	$2A'$	$A'+A''$	$A'+A''$	$2A_u$

† B species subscripts based on σ_h.

Correlation Tables (contd.): S_6, D_3, C_{3v}, D_{3d}, C_6, C_{3h}, C_{6h}, D_{3h}, C_{5h}, S_8, D_6, and C_{6v}.

S_6	C_3	C_i
A_g	A	A_g
E_g*	E*	$2A_g$
A_u	A	A_u
E_u*	E*	$2A_u$

D_3	C_3	C_2
A_1	A	A
A_2	A	B
E	E*	$A+B$

C_{3v}	C_3	C_s
A_1	A	A'
A_2	A	A''
E	E*	$A'+A''$

D_{3d}	D_3	C_{3v}	S_6	C_3	C_{2h}	C_2	C_s	C_i
A_{1g}	A_1	A_1	A_g	A	A_g	A	A'	A_g
A_{2g}	A_2	A_2	A_g	A	B_g	B	A''	A_g
E_g	E	E	E_g*	E*	A_g+B_g	$A+B$	$A'+A''$	$2A_g$
A_{1u}	A_1	A_2	A_u	A	A_u	A	A''	A_u
A_{2u}	A_2	A_1	A_u	A	B_u	B	A'	A_u
E_u	E	E	E_u*	E*	A_u+B_u	$A+B$	$A'+A''$	$2A_u$

C_6	C_3	C_2
A	A	A
B	A	B
E_1*	E*	$2B$
E_2*	E*	$2A$

C_{3h}	C_3	C_s
A'	A	A'
E'*	E*	$2A'$
A''	A	A''
E''*	E*	$2A''$

C_{6h}	C_6	C_{3h}	S_6	C_{2h}	C_3	C_2	C_s	C_i
A_g	A	A'	A_g	A_g	A	A	A'	A_g
B_g	B	A''	A_g	B_g	A	B	A''	A_g
E_{1g}*	E_1*	E''*	E_g*	$2B_g$	E*	$2B$	$2A''$	$2A_g$
E_{2g}*	E_2*	E'*	E_g*	$2A_g$	E*	$2A$	$2A'$	$2A_g$
A_u	A	A''	A_u	A_u	A	A	A''	A_u
B_u	B	A'	A_u	B_u	A	B	A'	A_u
E_{1u}*	E_1*	E'*	E_u*	$2B_u$	E*	$2B$	$2A'$	$2A_u$
E_{2u}*	E_2*	E''*	E_u*	$2A_u$	E*	$2A$	$2A''$	$2A_u$

D_{3h}	C_{3h}	D_3	C_{3v}	$\sigma_h \to \sigma'_v(yz)$ C_{2v}	C_3	C_2	σ_h C_s	σ_v C_s
A'_1	A'	A_1	A_1	A_1	A	A	A'	A'
A'_2	A'	A_2	A_2	B_2	A	B	A'	A''
E'	E'^*	E	E	A_1+B_2	E^*	$A+B$	$2A'$	$A'+A''$
A''_1	A''	A_1	A_2	A_2	A	A	A''	A''
A''_2	A''	A_2	A_1	B_1	A	B	A''	A'
E''	E''^*	E	E	A_2+B_1	E^*	$A+B$	$2A''$	$A'+A''$

C_{5h}	C_5	C_s
A'	A	A'
$E'_1{}^*$	$E_1{}^*$	$2A'$
$E'_2{}^*$	$E_2{}^*$	$2A'$
A''	A	A''
$E''_1{}^*$	$E_1{}^*$	$2A''$
$E''_2{}^*$	$E_2{}^*$	$2A''$

S_8	C_4	C_2
A	A	A
B	A	A
$E_1{}^*$	E^*	$2B$
$E_2{}^*$	$2B$	$2A$
$E_3{}^*$	E^*	$2B$

D_6	C_6	C'_2 D_3	C''_2 D_3	D_2	C_3	C_2	C'_2 C_2	C''_2 C_2
A_1	A	A_1	A_1	A	A	A	A	A
A_2	A	A_2	A_2	B_1	A	A	B	B
B_1	B	A_1	A_2	B_2	A	B	A	B
B_2	B	A_2	A_1	B_3	A	B	B	A
E_1	$E_1{}^*$	E	E	B_2+B_3	E^*	$2B$	$A+B$	$A+B$
E_2	$E_2{}^*$	E	E	$A+B_1$	E^*	$2A$	$A+B$	$A+B$

C_{6v}	C_6	σ_v C_{3v}	σ_d C_{3v}	$\sigma_v \to \sigma_v(xz)$ C_{2v}	C_3	C_2	σ_v C_s	σ_d C_s
A_1	A	A_1	A_1	A_1	A	A	A'	A'
A_2	A	A_2	A_2	A_2	A	A	A''	A''
B_1	B	A_1	A_2	B_1	A	B	A'	A''
B_2	B	A_2	A_1	B_2	A	B	A''	A'
E_1	$E_1{}^*$	E	E	B_1+B_2	E^*	$2B$	$A'+A''$	$A'+A''$
E_2	$E_2{}^*$	E	E	A_1+A_2	E^*	$2A$	$A'+A''$	$A'+A''$

Correlation Tables (contd.): D_{6h}.

D_{6h}	D_6	C_2' D_{3h}	C_2'' D_{3h}	C_{6v}†	C_{6h}	C_2'' D_{3d}	C_2' D_{3d}	$\sigma_h\to\sigma(xy)$ $\sigma_v\to\sigma(yz)$ D_{2h}	C_6	C_{3h}	C_2' D_3	C_2'' D_3	σ_v C_{3v}	σ_d C_{3v}	S_6	D_2‡	$\sigma_h\to\sigma_v(xz)$ C_2' C_{2v}
A_{1g}	A_1	A_1'	A_1'	A_1	A_g	A_{1g}	A_{1g}	A_g	A	A'	A_1	A_1	A_1	A_1	A_g	A	A_1
A_{2g}	A_2	A_2'	A_2'	A_2	A_g	A_{2g}	A_{2g}	B_{1g}	A	A'	A_2	A_2	A_2	A_2	A_g	B_1	B_1
B_{1g}	B_1	A_1''	A_2''	B_2	B_g	A_{2g}	A_{1g}	B_{2g}	B	A''	A_1	A_2	A_2	A_1	A_g	B_2	A_2
B_{2g}	B_2	A_2''	A_1''	B_1	B_g	A_{1g}	A_{2g}	B_{3g}	B	A''	A_2	A_1	A_1	A_2	A_g	B_3	B_2
E_{1g}	E_1	E''	E''	E_1	E_{1g}*	E_g	E_g	$B_{2g}+B_{3g}$	E_1*	E''*	E	E	E	E	E_g*	B_2+B_3	A_2+B_2
E_{2g}	E_2	E'	E'	E_2	E_{2g}*	E_g	E_g	A_g+B_{1g}	E_2*	E'*	E	E	E	E	E_g*	$A+B_1$	A_1+B_1
A_{1u}	A_1	A_1''	A_1''	A_2	A_u	A_{1u}	A_{1u}	A_u	A	A''	A_1	A_1	A_2	A_2	A_u	A	A_2
A_{2u}	A_2	A_2''	A_2''	A_1	A_u	A_{2u}	A_{2u}	B_{1u}	A	A''	A_2	A_2	A_1	A_1	A_u	B_1	B_2
B_{1u}	B_1	A_1'	A_2'	B_1	B_u	A_{2u}	A_{1u}	B_{2u}	B	A'	A_1	A_2	A_1	A_2	A_u	B_2	A_1
B_{2u}	B_2	A_2'	A_1'	B_2	B_u	A_{1u}	A_{2u}	B_{3u}	B	A'	A_2	A_1	A_2	A_1	A_u	B_3	B_1
E_{1u}	E_1	E'	E'	E_1	E_{1u}*	E_u	E_u	$B_{2u}+B_{3u}$	E_1*	E'*	E	E	E	E	E_u*	B_2+B_3	A_1+B_1
E_{2u}	E_2	E''	E''	E_2	E_{2u}*	E_u	E_u	A_u+B_{1u}	E_2*	E''*	E	E	E	E	E_u*	$A+B_1$	A_2+B_2

† $\sigma_v\to\sigma_v$; for $x\to x$ interchange B species subscripts.
‡ Cartesians retained.

$\sigma_h \rightarrow \sigma_v(xz)$

	C_2''	C_2	C_2'	C_2''		C_2	C_2'	C_2''	σ_h	σ_d	σ_v	
D_{6h}	C_{2v}	C_{2h}	C_{2h}	C_{2h}	C_3	C_2	C_2	C_2	C_s	C_s	C_s	C_i
A_{1g}	A_1	A_g	A_g	A_g	A	A	A	A	A'	A'	A'	A_g
A_{2g}	B_1	A_g	B_g	B_g	A	A	B	B	A'	A''	A''	A_g
B_{1g}	B_2	B_g	A_g	B_g	A	B	A	B	A''	A'	A''	A_g
B_{2g}	A_2	B_g	B_g	A_g	A	B	B	A	A''	A''	A'	A_g
E_{1g}	A_2+B_2	$2B_g$	A_g+B_g	A_g+B_g	E^*	$2B$	$A+B$	$A+B$	$2A''$	$A'+A''$	$A'+A''$	$2A_g$
E_{2g}	A_1+B_1	$2A_g$	A_g+B_g	A_g+B_g	E^*	$2A$	$A+B$	$A+B$	$2A'$	$A'+A''$	$A'+A''$	$2A_g$
A_{1u}	A_2	A_u	A_u	A_u	A	A	A	A	A''	A''	A''	A_u
A_{2u}	B_2	A_u	B_u	B_u	A	A	B	B	A''	A'	A'	A_u
B_{1u}	B_1	B_u	A_u	B_u	A	B	A	B	A'	A''	A'	A_u
B_{2u}	A_1	B_u	B_u	A_u	A	B	B	A	A'	A'	A''	A_u
E_{1u}	A_1+B_1	$2B_u$	A_u+B_u	A_u+B_u	E^*	$2B$	$A+B$	$A+B$	$2A'$	$A'+A''$	$A'+A''$	$2A_u$
E_{2u}	A_2+B_2	$2A_u$	A_u+B_u	A_u+B_u	E^*	$2A$	$A+B$	$A+B$	$2A''$	$A'+A''$	$A'+A''$	$2A_u$

Correlation Tables (contd.): D_5, C_{5v}, D_{5h}, D_{5d}, D_{4d}, D_{6d}, *and* $C_{\infty v}$.

D_5	C_5	C_2
A_1	A	A
A_2	A	B
E_1	$E_1{}^*$	$A+B$
E_2	$E_2{}^*$	$A+B$

C_{5v}	C_5	C_s
A_1	A	A'
A_2	A	A''
E_1	$E_1{}^*$	$A'+A''$
E_2	$E_2{}^*$	$A'+A''$

					$\sigma_h \rightarrow \sigma_v(xz)$		σ_h	σ_v
D_{5h}	D_5	C_{5v}	C_{5h}	C_5	C_{2v}	C_2	C_s	C_s
A_1'	A_1	A_1	A'	A	A_1	A	A'	A'
A_2	A_2	A_2	A'	A	B_1	B	A'	A''
E_1'	E_1	E_1	$E_1'{}^*$	$E_1{}^*$	A_1+B_1	$A+B$	$2A'$	$A'+A''$
E_2'	E_2	E_2	$E_2'{}^*$	$E_2{}^*$	A_1+B_1	$A+B$	$2A'$	$A'+A''$
A_1''	A_1	A_2	A''	A	A_2	A	A''	A''
A_2''	A_2	A_1	A''	A	B_2	B	A''	A'
E_1''	E_1	E_1	$E_1''{}^*$	$E_1{}^*$	A_2+B_2	$A+B$	$2A''$	$A'+A''$
E_2''	E_2	E_2	$E_2''{}^*$	$E_2{}^*$	A_2+B_2	$A+B$	$2A''$	$A'+A''$

D_{5d}	D_5	C_{5v}	C_5	C_2	C_s	C_i
A_{1g}	A_1	A_1	A	A	A'	A_g
A_{2g}	A_2	A_2	A	B	A''	A_g
E_{1g}	E_1	E_1	$E_1{}^*$	$A+B$	$A'+A''$	$2A_g$
E_{2g}	E_2	E_2	$E_2{}^*$	$A+B$	$A'+A''$	$2A_g$
A_{1u}	A_1	A_2	A	A	A''	A_u
A_{2u}	A_2	A_1	A	B	A'	A_u
E_{1u}	E_1	E_1	$E_1{}^*$	$A+B$	$A'+A''$	$2A_u$
E_{2u}	E_2	E_2	$E_2{}^*$	$A+B$	$A'+B''$	$2A_u$

						C_2	C_2'	
D_{4d}	D_4	C_{4v}	S_8	C_4	C_{2v}	C_2	C_2	C_s
A_1	A_1	A_1	A	A	A_1	A	A	A'
A_2	A_2	A_2	A	A	A_2	A	B	A''
B_1	A_1	A_2	B	A	A_2	A	A	A''
B_2	A_2	A_1	B	A	A_1	A	B	A'
E_1	E	E	$E_1{}^*$	E^*	B_1+B_2	$2B$	$A+B$	$A'+A''$
E_2	B_1+B_2	B_1+B_2	$E_2{}^*$	$2B$	A_1+A_2	$2A$	$A+B$	$A'+A''$
E_3	E	E	$E_3{}^*$	E^*	B_1+B_2	$2B$	$A+B$	$A'+A''$

D_{6d}	D_6	C_{6v}	C_6	D_{2d}	D_3	C_{3v}	D_2	C_{2v}	S_4	C_3	C_2 (C_2)	C_2' (C_2)	C_s
A_1	A_1	A_1	A	A_1	A_1	A_1	A	A_1	A	A	A	A	A'
A_2	A_2	A_2	A	A_2	A_2	A_2	B_1	A_2	A	A	A	B	A''
B_1	A_1	A_2	A	B_1	A_1	A_2	A	A_2	B	A	A	A	A''
B_2	A_2	A_1	A	B_2	A_2	A_1	B_1	A_1	B	A	A	B	A'
E_1	E_1	E_1	$E_1{}^*$	E	E	E	B_2+B_3	B_1+B_2	E^*	E^*	$2B$	$A+B$	$A'+A''$
E_2	E_2	E_2	$E_2{}^*$	B_1+B_2	E	E	$A+B_1$	A_1+A_2	$2B$	E^*	$2A$	$A+B$	$A'+A''$
E_3	B_1+B_2	B_1+B_2	$2B$	E	A_1+A_2	A_1+A_2	B_2+B_3	B_1+B_2	E^*	$2A$	$2B$	$A+B$	$A'+A''$
E_4	E_2	E_2	$E_2{}^*$	A_1+A_2	E	E	$A+B_1$	A_1+A_2	$2A$	E^*	$2A$	$A+B$	$A'+A''$
E_5	E_1	E_1	$E_1{}^*$	E	E	E	B_2+B_3	B_1+B_2	E^*	E^*	$2B$	$A+B$	$A'+A''$

$C_{\infty v}$	C_{7v}	C_{6v}	C_{5v}	C_{4v}	C_{3v}	C_{2v}	C_s	$(\equiv C_{1v})$
Σ^+	A_1	A_1	A_1	A_1	A_1	A_1	A'	(A_1)
Σ^-	A_2	A_2	A_2	A_2	A_2	A_2	A''	(A_2)
Π	E_1	E_1	E_1	E	E	B_1+B_2	$A'+A''$	(A_1+A_2)
Δ	E_2	E_2	E_2	B_1+B_2	E	A_1+A_2	$A'+A''$	$\vdots$
Φ	E_3	B_1+B_2	E_2	E	A_1+A_2	B_1+B_2	$A'+A''$	
$\vdots$	E_3	E_2	E_1	A_1+A_2	$\vdots$	$\vdots$	$\vdots$	
	E_2	E_1	A_1+A_2	$\vdots$				
	E_1	A_1+A_2	$\vdots$					
	A_1+A_2	$\vdots$						
	$\vdots$							

Correlation Tables (contd.): $D_{\infty h}$, T, T_h, *and* T_d.

$D_{\infty h}/D_{nh}$ (n even) or D_{nd} (n odd) as $C_{\infty v}/C_{nv}$ but with g/u rule. Note "D_{1d}" $\equiv C_{2h}$. Nomenclature rules also require interchange of subscripts for non-degenerate *ungerade* species of D_{nh} and D_{nd}. C_{2h} (D_{1d}) equivalents are: A_g (A_{1g}), B_g (A_{2g}), A_u (A_{2u}), B_u (A_{1u}) *after* interchange of subscripts.

$D_{\infty h}$	D_{7h}	D_{6d}	D_{5h}	D_{4d}	D_{3h}	D_{2d}	$\sigma_v \rightarrow \sigma_v(xz)$ C_{2v}
Σ_g^+	A_1'	A_1	A_1'	A_1	A_1'	A_1	A_1
Σ_g^-	A_2'	A_2	A_2'	A_2	A_2'	A_2	B_2
Π_g	E_1''	E_5	E_1''	E_3	E''	E	A_2+B_1
Δ_g	E_2'	E_2	E_2'	E_2	E'	B_1+B_2	A_1+B_2
Φ_g	E_3''	E_3	E_2''	E_1	$A_1''+A_2''$	E	A_2+B_1
⋮	E_3'	E_4	E_1'	B_1+B_2	E'	A_1+A_2	⋮
	E_2''	E_1	$A_1''+A_2''$	⋮	⋮	⋮	
	E_1'	B_1+B_2	⋮				
	$A_1''+A_2''$	⋮					
	⋮						
Σ_u^+	A_2''	B_2	A_2''	B_2	A_2''	B_2	B_1
Σ_u^-	A_1''	B_1	A_1''	B_1	A_1''	B_1	A_2
Π_u	E_1'	E_1	E_1'	E_1	E'	E	A_1+B_2
Δ_u	E_2''	E_4	E_2''	E_2	E''	A_1+A_2	A_2+B_1
Φ_u	E_3'	E_3	E_2'	E_3	$A_1'+A_2'$	E	A_1+B_2
⋮	E_3''	E_2	E_1''	A_1+A_2	E''	B_1+B_2	⋮
	E_2'	E_5	$A_1'+A_2'$	⋮	⋮	⋮	
	E_1''	A_1+A_2	⋮				
	$A_1'+A_2'$	⋮					
	⋮						

T	D_2	C_3	C_2
A	A	A	A
E^*	$2A$	E^*	$2A$
F	$B_1+B_2+B_3$	$A+E^*$	$A+2B$

T_h	T	D_{2h}	S_6	D_2	C_{2v}	C_{2h}	C_3	C_2	C_s	C_i
A_g	A	A_g	A_g	A	A_1	A_g	A	A	A'	A_g
$E_g{}^*$	E^*	$2A_g$	$E_g{}^*$	$2A$	$2A_1$	$2A_g$	E^*	$2A$	$2A'$	$2A_g$
F_g	F	$B_{1g}+B_{2g}+B_{3g}$	$A_g+E_g{}^*$	$B_1+B_2+B_3$	$A_2+B_1+B_2$	A_g+2B_g	$A+E^*$	$A+2B$	$A'+2A''$	$3A_g$
A_u	A	A_u	A_u	A	A_2	A_u	A	A	A''	A_u
$E_u{}^*$	E^*	$2A_u$	$E_u{}^*$	$2A$	$2A_2$	$2A_u$	E^*	$2A$	$2A''$	$2A_u$
F_u	F	$B_{1u}+B_{2u}+B_{3u}$	$A_u+E_u{}^*$	$B_1+B_2+B_3$	$A_1+B_1+B_2$	A_u+2B_u	$A+E^*$	$A+2B$	$2A'+A''$	$3A_u$

T_d	T	D_{2d}	C_{3v}	S_4	D_2	C_{2v}	C_3	C_2	C_s
A_1	A	A_1	A_1	A	A	A_1	A	A	A'
A_2	A	B_1	A_2	B	A	A_2	A	A	A''
E	E^*	A_1+B_1	E	$A+B$	$2A$	A_1+A_2	E^*	$2A$	$A'+A''$
F_1	F	A_2+E	A_2+E	$A+E^*$	$B_1+B_2+B_3$	$A_2+B_1+B_2$	$A+E^*$	$A+2B$	$A'+2A''$
F_2	F	B_2+E	A_1+E	$B+E^*$	$B_1+B_2+B_3$	$A_1+B_1+B_2$	$A+E^*$	$A+2B$	$2A'+A''$

Correlation Tables (contd.): O, O_h, *and* I_h.

O	T	D_4	D_3	C_4	$3C_2$ D_2	$C_2, 2C'_2$ D_2	C_3	C_2	C_2
A_1	A	A_1	A_1	A	A	A	A	A	A
A_2	A	B_1	A_2	B	A	B_1	A	A	B
E	E^*	A_1+B_1	E	$A+B$	$2A$	$A+B_1$	E^*	$2A$	$A+B$
F_1	F	A_2+E	A_2+E	$A+E^*$	$B_1+B_2+B_3$	$B_1+B_2+B_3$	$A+E^*$	$A+2B$	$A+2B$
F_2	F	B_2+E	A_1+E	$B+E^*$	$B_1+B_2+B_3$	$A+B_2+B_3$	$A+E^*$	$A+2B$	$2A+B$

O_h	O	T_d	T_h	D_{4h}	D_{3d}
A_{1g}	A_1	A_1	A_g	A_{1g}	A_{1g}
A_{2g}	A_2	A_2	A_g	B_{1g}	A_{2g}
E_g	E	E	E_g^*	$A_{1g}+B_{1g}$	E_g
F_{1g}	F_1	F_1	F_g	$A_{2g}+E_g$	$A_{2g}+E_g$
F_{2g}	F_2	F_2	F_g	$B_{2g}+E_g$	$A_{1g}+E_g$
A_{1u}	A_1	A_2	A_u	A_{1u}	A_{1u}
A_{2u}	A_2	A_1	A_u	B_{1u}	A_{2u}
E_u	E	E	E_u^*	$A_{1u}+B_{1u}$	E_u
F_{1u}	F_1	F_2	F_u	$A_{2u}+E_u$	$A_{2u}+E_u$
F_{2u}	F_2	F_1	F_u	$B_{2u}+E_u$	$A_{1u}+E_u$

I_h	D_{5d}	D_{3d}	T_h
A_g	A_{1g}	A_{1g}	A_g
F_{1g}	$A_{2g}+E_{1g}$	$A_{2g}+E_g$	F_g
F_{2g}	$A_{2g}+E_{2g}$	$A_{2g}+E_g$	F_g
G_g	$E_{1g}+E_{2g}$	$A_{1g}+A_{2g}+E_g$	A_g+F_g
H_g	$A_{1g}+E_{1g}+E_{2g}$	$A_{1g}+2E_g$	$E_g{}^*+F_g$
A_u	A_{1u}	A_{1u}	A_u
F_{1u}	$A_{2u}+E_{1u}$	$A_{2u}+E_u$	F_u
F_{2u}	$A_{2u}+E_{2u}$	$A_{2u}+E_u$	F_u
G_u	$E_{1u}+E_{2u}$	$A_{1u}+A_{2u}+E_u$	A_u+F_u
H_u	$A_{1u}+E_{1u}+E_{2u}$	$A_{1u}+2E_u$	$E_u{}^*+F_u$

A2.4 Possible Site Symmetries (atoms and molecules)

This table is necessarily very similar to those given by Halford (see bibliography to Chapter 4) and Adams (see Chapter 9). It is a summary of information contained in *International Tables* (see bibliography to Chapter 4). Only *point* sites are included, and only symmetries greater than C_1. A table of Plane and Line sites would be very cumbersome, but it may be noted that the *minimum* requirement is that the Plane or Line group point symmetry should be either the same as the Space group point symmetry or a subgroup of it.

In the table below, the point group is preceded by the number of distinct sets of sites of that symmetry, and followed by the number per set in brackets. In the case of point groups without a fixed origin (C_s, C_n, C_{nv}), the listed sites may in principle contain any number of discrete molecules, atoms, or ions of that symmetry. (This is formally true of molecules with other point groups, but where the origin is defined, a second molecule would either have to interlock with or encage the first, a much less likely situation.)

The number of sites per set is given for the conventional unit cell, and should be divided by the cell multiplicity to obtain the factor group correlation number (i.e. the sites per primitive cell). This is indicated in square brackets at the end of each list; e.g. [÷4] for an *F*-cell.

Space group		Site symmetries
2 $P\bar{1}$	C_i^1	$8C_i$
3 $P2$	C_2^1	$4C_2$
5 $C2$	C_2^3	$2C_2(2)$ [÷2]

6	*Pm*	C_s^1	$2C_s$
8	*Cm*	C_s^3	$C_s(2)$ [÷2]
10	*P2/m*	C_{2h}^1	$8C_{2h}$; $4C_2(2)$; $2C_s(2)$
11	$P2_1/m$	C_{2h}^2	$4C_i(2)$; $C_s(2)$
12	*C2/m*	C_{2h}^3	$4C_{2h}(2)$; $2C_i(4)$; $2C_2(4)$; $C_s(4)$ [÷2]
13	*P2/c*	C_{2h}^4	$4C_i(2)$; $2C_2(2)$
14	$P2_1/c$	C_{2h}^5	$4C_i(2)$
15	*C2/c*	C_{2h}^6	$4C_i(4)$; $C_2(4)$ [÷2]
16	*P222*	D_2^1	$8D_2$; $12C_2(2)$
17	$P222_1$	D_2^2	$4C_2(2)$
18	$P2_12_12$	D_2^3	$2C_2(2)$
20	$C222_1$	D_2^5	$2C_2(4)$ [÷2]
21	*C222*	D_2^6	$4D_2(2)$; $7C_2(4)$ [÷2]
22	*F222*	D_2^7	$4D_2(4)$; $6C_2(8)$ [÷4]
23	*I222*	D_2^8	$4D_2(2)$; $6C_2(4)$ [÷2]
24	$I2_12_12_1$	D_2^9	$3C_2(4)$ [÷2]
25	*Pmm2*	C_{2v}^1	$4C_{2v}$; $4C_s(2)$
26	$Pmc2_1$	C_{2v}^2	$2C_s(2)$
27	*Pcc2*	C_{2v}^3	$4C_2(2)$
28	*Pma2*	C_{2v}^4	$2C_2(2)$; $C_s(2)$
30	*Pnc2*	C_{2v}^6	$2C_2(2)$
31	$Pmn2_1$	C_{2v}^7	$C_s(2)$
32	*Pba2*	C_{2v}^8	$2C_2(2)$
34	*Pnn2*	C_{2v}^{10}	$2C_2(2)$
35	*Cmm2*	C_{2v}^{11}	$2C_{2v}(2)$; $C_2(4)$; $2C_s(4)$ [÷2]
36	$Cmc2_1$	C_{2v}^{12}	$C_s(4)$ [÷2]
37	*Ccc2*	C_{2v}^{13}	$3C_2(4)$ [÷2]
38	*Amm2*	C_{2v}^{14}	$2C_{2v}(2)$; $3C_s(4)$ [÷2]
39	*Abm2*	C_{2v}^{15}	$2C_2(4)$; $C_s(4)$ [÷2]
40	*Ama2*	C_{2v}^{16}	$C_2(4)$; $C_s(4)$ [÷2]
41	*Aba2*	C_{2v}^{17}	$C_2(4)$ [÷2]
42	*Fmm2*	C_{2v}^{18}	$C_{2v}(4)$; $C_2(8)$; $2C_s(8)$ [÷4]
43	*Fdd2*	C_{2v}^{19}	$C_2(8)$ [÷4]
44	*Imm2*	C_{2v}^{20}	$2C_{2v}(2)$; $2C_s(4)$ [÷2]
45	*Iba2*	C_{2v}^{21}	$2C_2(4)$ [÷2]
46	*Ima2*	C_{2v}^{22}	$C_2(4)$; $C_s(4)$ [÷2]
47	*Pmmm*	D_{2h}^1	$8D_{2h}$; $12C_{2v}(2)$; $6C_s(4)$
48	*Pnnn*	D_{2h}^2	$4D_2(2)$; $2C_i(4)$; $6C_2(4)$
49	*Pccm*	D_{2h}^3	$4C_{2h}(2)$; $4D_2(2)$; $8C_2(4)$; $C_s(4)$
50	*Pban*	D_{2h}^4	$4D_2(2)$; $2C_i(4)$; $6C_2(4)$

51	*Pmma*	D_{2h}^5	$4C_{2h}(2)$; $2C_{2v}(2)$; $2C_2(4)$; $3C_s(4)$
52	*Pnna*	D_{2h}^6	$2C_i(4)$; $2C_2(4)$
53	*Pmna*	D_{2h}^7	$4C_{2h}(2)$; $3C_2(4)$; $C_s(4)$
54	*Pcca*	D_{2h}^8	$2C_i(4)$; $3C_2(4)$
55	*Pbam*	D_{2h}^9	$4C_{2h}(2)$; $2C_2(4)$; $2C_s(4)$
56	*Pccn*	D_{2h}^{10}	$2C_i(4)$; $2C_2(4)$
57	*Pbcm*	D_{2h}^{11}	$2C_i(4)$; $C_2(4)$; $C_s(4)$
58	*Pnnm*	D_{2h}^{12}	$4C_{2h}(2)$; $2C_2(4)$; $C_s(4)$
59	*Pmmn*	D_{2h}^{13}	$2C_{2v}(2)$; $2C_i(4)$; $2C_s(4)$
60	*Pbcn*	D_{2h}^{14}	$2C_i(4)$; $C_2(4)$
61	*Pbca*	D_{2h}^{15}	$2C_i(4)$
62	*Pnma*	D_{2h}^{16}	$2C_i(4)$; $C_s(4)$
63	*Cmcm*	D_{2h}^{17}	$2C_{2h}(4)$; $C_{2v}(4)$; $C_i(8)$; $C_2(8)$; $2C_s(8)$ [÷2]
64	*Cmca*	D_{2h}^{18}	$2C_{2h}(4)$; $C_i(8)$; $C_s(8)$; $2C_2(8)$ [÷2]
65	*Cmmm*	D_{2h}^{19}	$4D_{2h}(2)$; $2C_{2h}(4)$; $6C_{2v}(4)$; $C_2(8)$; $4C_s(8)$ [÷2]
66	*Cccm*	D_{2h}^{20}	$2D_2(4)$; $4C_{2h}(4)$; $5C_2(8)$; $C_s(8)$ [÷2]
67	*Cmma*	D_{2h}^{21}	$2D_2(4)$; $4C_{2h}(4)$; $C_{2v}(4)$; $5C_2(8)$; $2C_s(8)$ [÷2]
68	*Ccca*	D_{2h}^{22}	$2D_2(4)$; $2C_i(8)$; $4C_2(8)$ [÷2]
69	*Fmmm*	D_{2h}^{23}	$2D_{2h}(4)$; $3C_{2h}(8)$; $D_2(8)$; $3C_{2v}(8)$; $3C_2(16)$; $3C_s(16)$ [÷4]
70	*Fddd*	D_{2h}^{24}	$2D_2(8)$; $2C_i(16)$; $3C_2(16)$ [÷4]
71	*Immm*	D_{2h}^{25}	$4D_{2h}(2)$; $6C_{2v}(4)$; $C_i(8)$; $3C_s(8)$ [÷2]
72	*Ibam*	D_{2h}^{26}	$2D_2(4)$; $2C_{2h}(4)$; $C_i(8)$; $4C_2(8)$; $C_s(8)$ [÷2]
73	*Ibca*	D_{2h}^{27}	$2C_i(8)$; $3C_2(8)$; [÷2]
74	*Imma*	D_{2h}^{28}	$4C_{2h}(4)$; $C_{2v}(4)$; $2C_2(8)$; $2C_s(8)$ [÷2]
75	*P*4	C_4^1	$2C_4$; $C_2(2)$
77	*P*4$_2$	C_4^3	$3C_2(2)$
79	*I*4	C_4^5	$C_4(2)$; $C_2(4)$ [÷2]
80	*I*4$_1$	C_4^6	$C_2(4)$ [÷2]
81	*P*$\bar{4}$	S_4^1	$4S_4$; $3C_2(2)$
82	*I*$\bar{4}$	S_4^2	$4S_4(2)$; $2C_2(4)$ [÷2]
83	*P*4/*m*	C_{4h}^1	$4C_{4h}$; $2C_{2h}(2)$; $2C_4(2)$; $C_2(4)$; $2C_s(4)$
84	*P*4$_2$/*m*	C_{4h}^2	$4C_{2h}(2)$; $2S_4(2)$; $3C_2(4)$; $C_s(4)$
85	*P*4/*n*	C_{4h}^3	$2S_4(2)$; $C_4(2)$; $2C_i(4)$; $C_2(4)$
86	*P*4$_2$/*n*	C_{4h}^4	$2S_4(2)$; $2C_i(4)$; $2C_2(4)$
87	*I*4/*m*	C_{4h}^5	$2C_{4h}(2)$; $C_{2h}(4)$; $S_4(4)$; $C_4(4)$; $C_i(8)$; $C_2(8)$; $C_s(8)$ [÷2]
88	*I*4$_1$/*a*	C_{4h}^6	$2S_4(4)$; $2C_i(8)$; $C_2(8)$ [÷2]
89	*P*422	D_4^1	$4D_4$; $2D_2(2)$; $2C_4(2)$; $7C_2(4)$
90	*P*42$_1$2	D_4^2	$2D_2(2)$; $C_4(2)$; $3C_2(4)$

91	$P4_122$	D_4^3	$3C_2(4)$
92	$P4_12_12$	D_4^4	$C_2(4)$
93	$P4_222$	D_4^5	$6D_2(2)$; $9C_2(4)$
94	$P4_22_12$	D_4^6	$2D_2(2)$; $4C_2(4)$
95	$P4_322$	D_4^7	$3C_2(4)$
96	$P4_32_12$	D_4^8	$C2(4)$
97	$I422$	D_4^9	$2D_4(2)$; $2D_2(4)$; $C_4(4)$; $5C_2(8)$ [÷2]
98	$I4_122$	D_4^{10}	$2D_2(4)$; $4C_2(8)$ [÷2]
99	$P4mm$	C_{4v}^1	$2C_{4v}$; $C_{2v}(2)$; $3C_s(4)$
100	$P4bm$	C_{4v}^2	$C_4(2)$; $C_{2v}(2)$; $C_s(4)$
101	$P4_2cm$	C_{4v}^3	$2C_{2v}(2)$; $C_2(4)$; $C_s(4)$
102	$P4_2nm$	C_{4v}^4	$C_{2v}(2)$; $C_2(4)$; $C_s(4)$
103	$P4cc$	C_{4v}^5	$2C_4(2)$; $C_2(4)$
104	$P4nc$	C_{4v}^6	$C_4(2)$; $C_2(4)$
105	$P4_2mc$	C_{4v}^7	$3C_{2v}(2)$; $2C_s(4)$
106	$P4_2bc$	C_{4v}^8	$2C_2(4)$
107	$I4mm$	C_{4v}^9	$C_{4v}(2)$; $C_{2v}(4)$; $2C_s(8)$ [÷2]
108	$I4cm$	C_{4v}^{10}	$C_4(4)$; $C_{2v}(4)$; $C_s(8)$ [÷2]
109	$I4_1md$	C_{4v}^{11}	$C_{2v}(4)$; $C_s(8)$ [÷2]
110	$I4_1cd$	C_{4v}^{12}	$C_2(8)$ [÷2]
111	$P\bar{4}2m$	D_{2d}^1	$4D_{2d}$; $2D_2(2)$; $2C_{2v}(2)$; $5C_2(4)$; $C_s(4)$
112	$P\bar{4}2c$	D_{2d}^2	$4D_2(2)$; $2S_4(2)$; $7C_2(4)$
113	$P\bar{4}2_1m$	D_{2d}^3	$2S_4(2)$; $C_{2v}(2)$; $C_2(4)$; $C_s(4)$
114	$P\bar{4}2_1c$	D_{2d}^4	$2S_4(2)$; $2C_2(4)$
115	$P\bar{4}m2$	D_{2d}^5	$4D_{2d}$; $3C_{2v}(2)$; $2C_2(4)$; $2C_s(4)$
116	$P\bar{4}c2$	D_{2d}^6	$2D_2(2)$; $2S_4(2)$; $5C_2(4)$
117	$P\bar{4}b2$	D_{2d}^7	$2D_2(2)$; $2S_4(2)$; $4C_2(4)$
118	$P\bar{4}n2$	D_{2d}^8	$2D_2(2)$; $2S_4(2)$; $4C_2(4)$
119	$I\bar{4}m2$	D_{2d}^9	$4D_{2d}(2)$; $2C_{2v}(4)$; $2C_2(8)$; $C_s(8)$ [÷2]
120	$I\bar{4}c2$	D_{2d}^{10}	$2D_2(4)$; $2S_4(4)$; $4C_2(8)$ [÷2]
121	$I\bar{4}2m$	D_{2d}^{11}	$2D_{2d}(2)$; $D_2(4)$; $S_4(4)$; $C_{2v}(4)$; $3C_2(8)$; $C_s(8)$ [÷2]
122	$I\bar{4}2d$	D_{2d}^{12}	$2S_4(4)$; $2C_2(8)$ [÷2]
123	$P4/mmm$	D_{4h}^1	$4D_{4h}$; $2D_{2h}(2)$; $2C_{4v}(2)$; $7C_{2v}(4)$; $5C_s(8)$
124	$P4/mcc$	D_{4h}^2	$2D_4(2)$; $2C_{4h}(2)$; $D_2(4)$; $C_{2h}(4)$; $2C_4(4)$; $4C_2(8)$; $C_s(8)$
125	$P4/nbm$	D_{4h}^3	$2D_4(2)$; $2D_{2d}(2)$; $2C_{2h}(4)$; $C_4(4)$; $C_{2v}(4)$; $4C_2(8)$; $C_s(8)$
126	$P4/nnc$	D_{4h}^4	$2D_4(2)$; $D_2(4)$; $S_4(4)$; $C_4(4)$; $C_i(8)$; $4C_2(8)$
127	$P4/mbm$	D_{4h}^5	$2C_{4h}(2)$; $2D_{2h}(2)$; $C_4(4)$; $3C_{2v}(4)$; $3C_s(8)$
128	$P4/mnc$	D_{4h}^6	$2C_{4h}(2)$; $C_{2h}(4)$; $D_2(4)$; $C_4(4)$; $2C_2(8)$; $C_s(8)$

129	$P4/nmm$	D_{4h}^{7}	$2D_{2d}(2)$; $C_{4v}(2)$; $2C_{2h}(4)$; $C_{2v}(4)$; $2C_2(8)$; $2C_s(8)$
130	$P4/ncc$	D_{4h}^{8}	$D_2(4)$; $S_4(4)$; $C_4(4)$; $C_i(8)$; $2C_2(8)$
131	$P4_2/mmc$	D_{4h}^{9}	$4D_{2h}(2)$; $2D_{2d}(2)$; $7C_{2v}(4)$; $C_2(8)$; $3C_s(8)$
132	$P4_2/mcm$	D_{4h}^{10}	$2D_{2h}(2)$; $2D_{2d}(2)$; $D_2(4)$; $C_{2h}(4)$; $4C_{2v}(4)$; $3C_2(8)$; $2C_s(8)$
133	$P4_2/nbc$	D_{4h}^{11}	$3D_2(4)$; $S_4(4)$; $C_i(8)$; $5C_2(8)$
134	$P4_2/nnm$	D_{4h}^{12}	$2D_{2d}(2)$; $2D_2(4)$; $2C_{2h}(4)$; $C_{2v}(4)$; $5C_2(8)$; $C_s(8)$
135	$P4_2/mbc$	D_{4h}^{13}	$2C_{2h}(4)$; $S_4(4)$; $D_2(4)$; $3C_2(8)$; $C_s(8)$
136	$P4_2/mnm$	D_{4h}^{14}	$2D_{2h}(2)$; $C_{2h}(4)$; $S_4(4)$; $3C_{2v}(4)$; $C_2(8)$; $2C_s(8)$
137	$P4_2/nmc$	D_{4h}^{15}	$2D_{2d}(2)$; $2C_{2v}(4)$; $C_i(8)$; $C_2(8)$; $C_s(8)$
138	$P4_2/ncm$	D_{4h}^{16}	$D_2(4)$; $S_4(4)$; $2C_{2h}(4)$; $C_{2v}(4)$; $3C_2(8)$; $C_s(8)$
139	$I4/mmm$	D_{4h}^{17}	$2D_{4h}(2)$; $D_{2h}(4)$; $D_{2d}(4)$; $C_{4v}(4)$; $C_{2h}(8)$; $4C_{2v}(8)$; $C_2(16)$; $3C_s(16)$ [÷2]
140	$I4/mcm$	D_{4h}^{18}	$D_4(4)$; $D_{2d}(4)$; $C_{4h}(4)$; $D_{2h}(4)$; $C_{2h}(8)$; $C_4(8)$; $2C_{2v}(8)$; $2C_2(16)$; $2C_s(16)$ [÷2]
141	$I4_1/amd$	D_{4h}^{19}	$2D_{2d}(4)$; $2C_{2h}(8)$; $C_{2v}(8)$; $2C_2(16)$; $C_s(16)$ [÷2]
142	$I4_1/acd$	D_{4h}^{20}	$S_4(8)$; $D_2(8)$; $C_i(16)$; $3C_2(16)$ [÷2]
143	$P3$	C_3^1	$3C_3$
146	$R3$	C_3^4	C_3
147	$P\bar{3}$	C_{3i}^1	$2C_{3i}$; $2C_3(2)$; $2C_i(3)$
148	$R\bar{3}$	C_{3i}^2	$2C_{3i}$; $C_3(2)$; $2C_i(3)$
149	$P312$	D_3^1	$6D_3$; $3C_3(2)$; $2C_2(3)$
150	$P321$	D_3^2	$2D_3$; $2C_3(2)$; $2C_2(3)$
151	$P3_112$	D_3^3	$2C_2(3)$
152	$P3_121$	D_3^4	$2C_2(3)$
153	$P3_212$	D_3^5	$2C_2(3)$
154	$P3_221$	D_3^6	$2C_2(3)$
155	$R32$	D_3^7	$2D_3$; $C_3(2)$; $2C_2(3)$
156	$P3m1$	C_{3v}^1	$3C_{3v}$; $C_s(3)$
157	$P31m$	C_{3v}^2	C_{3v}; $C_3(2)$; $C_s(3)$
158	$P3c1$	C_{3v}^3	$3C_3(2)$
159	$P31c$	C_{3v}^4	$2C_3(2)$
160	$R3m$	C_{3v}^5	C_{3v}; $C_s(3)$
161	$R3c$	C_{3v}^6	$C_3(2)$
162	$P\bar{3}1m$	D_{3d}^1	$2D_{3d}$; $2D_3(2)$; $C_{3v}(2)$; $2C_{2h}(3)$; $C_3(4)$; $2C_2(6)$; $C_s(6)$
163	$P\bar{3}1c$	D_{3d}^2	$3D_3(2)$; $C_{3i}(2)$; $2C_3(4)$; $C_i(6)$; $C_2(6)$
164	$P\bar{3}m1$	D_{3d}^3	$2D_{3d}$; $2C_{3v}(2)$; $2C_{2h}(3)$; $2C_2(6)$; $C_s(6)$
165	$P\bar{3}c1$	D_{3d}^4	$D_3(2)$; $C_{3i}(2)$; $2C_3(4)$; $C_i(6)$; $C_2(6)$
166	$R3m$	D_{3d}^5	$2D_{3d}$; $C_{3v}(2)$; $2C_{2h}(3)$; $2C_2(6)$; $C_s(6)$

167	$R\bar{3}c$	D_{3d}^6	$D_3(2)$; $C_{3i}(2)$; $C_3(4)$; $C_i(6)$; $C_2(6)$
168	$P6$	C_6^1	C_6; $C_3(2)$; $C_2(3)$
171	$P6_2$	C_6^4	$2C_2(3)$
172	$P6_4$	C_6^5	$2C_2(3)$
173	$P6_3$	C_6^6	$2C_3(2)$
174	$P\bar{6}$	C_{3h}^1	$6C_{3h}$; $3C_3(2)$; $2C_s(3)$
175	$P6/m$	C_{6h}^1	$2C_{6h}$; $2C_{3h}(2)$; $C_6(2)$; $2C_{2h}(3)$; $C_3(4)$; $C_2(6)$; $2C_s(6)$
176	$P6_3/m$	C_{6h}^2	$3C_{3h}(2)$; $C_{3i}(2)$; $2C_3(4)$; $C_i(6)$; $C_s(6)$
177	$P622$	D_6^1	$2D_6$; $2D_3(2)$; $C_6(2)$; $2D_2(3)$; $C_3(4)$; $5C_2(6)$
178	$P6_122$	D_6^2	$2C_2(6)$
179	$P6_522$	D_6^3	$2C_2(6)$
180	$P6_222$	D_6^4	$4D_2(3)$; $6C_2(6)$
181	$P6_422$	D_6^5	$4D_2(3)$; $6C_2(6)$
182	$P6_322$	D_6^6	$4D_3(2)$; $2C_3(4)$; $2C_2(6)$
183	$P6mm$	C_{6v}^1	C_{6v}; $C_{3v}(2)$; $C_{2v}(3)$; $2C_s(6)$
184	$P6cc$	C_{6v}^2	$C_6(2)$; $C_3(4)$; $C_2(6)$
185	$P6_3cm$	C_{6v}^3	$C_{3v}(2)$; $C_3(4)$; $C_s(6)$
186	$P6_3mc$	C_{6v}^4	$2C_{3v}(2)$; $C_s(6)$
187	$P\bar{6}m2$	D_{3h}^1	$6D_{3h}$; $3C_{3v}(2)$; $2C_{2v}(3)$; $3C_s(6)$
188	$P\bar{6}c2$	D_{3h}^2	$3D_3(2)$; $3C_{3h}(2)$; $3C_3(4)$; $C_2(6)$; $C_s(6)$
189	$P\bar{6}2m$	D_{3h}^3	$2D_{3h}$; $2C_{3h}(2)$; $C_{3v}(2)$; $2C_{2v}(3)$; $C_3(4)$; $3C_s(6)$
190	$P\bar{6}2c$	D_{3h}^4	$D_3(2)$; $3C_{3h}(2)$; $2C_3(4)$; $C_2(6)$; $C_s(6)$
191	$P6/mmm$	D_{6h}^1	$2D_{6h}$; $2D_{3h}(2)$; $C_{6v}(2)$; $2D_{2h}(3)$; $C_{3v}(4)$; $5C_{2v}(6)$; $4C_s(12)$
192	$P6/mcc$	D_{6h}^2	$D_6(2)$; $C_{6h}(2)$; $D_3(4)$; $C_{3h}(4)$; $C_6(4)$; $D_2(6)$; $C_{2h}(6)$; $C_3(8)$; $3C_2(12)$; $C_s(12)$
193	$P6_3/mcm$	D_{6h}^3	$D_{3h}(2)$; $D_{3d}(2)$; $C_{3h}(4)$; $D_3(4)$; $C_{3v}(4)$; $C_{2h}(6)$; $C_{2v}(6)$; $C_3(8)$; $C_2(12)$; $2C_s(12)$
194	$P6_3/mmc$	D_{6h}^4	$D_{3d}(2)$; $3D_{3h}(2)$; $2C_{3v}(4)$; $C_{2h}(6)$; $C_{2v}(6)$; $C_2(12)$; $2C_s(12)$
195	$P23$	T^1	$2T$; $2D_2(3)$; $C_3(4)$; $4C_2(6)$
196	$F23$	T^2	$4T(4)$; $C_3(16)$; $2C_2(24)$ [$\div 4$]
197	$I23$	T^3	$T(2)$; $D_2(6)$; $C_3(8)$; $2C_2(12)$ [$\div 2$]
198	$P2_13$	T^4	$C_3(4)$
199	$I2_13$	T^5	$C_3(8)$; $C_2(12)$ [$\div 2$]
200	$Pm3$	T_h^1	$2T_h$; $2D_{2h}(3)$; $4C_{2v}(6)$; $C_3(8)$; $2C_s(12)$
201	$Pn3$	T_h^2	$T(2)$; $2C_{3i}(4)$; $D_2(6)$; $C_3(8)$; $2C_2(12)$
202	$Fm3$	T_h^3	$2T_h(4)$; $T(8)$; $C_{2h}(24)$; $C_{2v}(24)$; $C_3(32)$; $C_2(48)$; $C_s(48)$ [$\div 4$]

203	$Fd3$	T_h^4	$2T(8)$; $2C_{3i}(16)$; $C_3(32)$; $C_2(48)$ $[\div 4]$
204	$Im3$	T_h^5	$T_h(2)$; $D_{2h}(6)$; $C_{3i}(8)$; $2C_{2v}(12)$; $C_3(16)$; $C_s(24)$ $[\div 2]$
205	$Pa3$	T_h^6	$2C_{3i}(4)$; $C_3(8)$
206	$Ia3$	T_h^7	$2C_{3i}(8)$; $C_3(16)$; $C_2(24)$ $[\div 2]$
207	$P432$	O^1	$2O$; $2D_4(3)$; $2C_4(6)$; $C_3(8)$; $3C_2(12)$
208	$P4_232$	O^2	$T(2)$; $2D_3(4)$; $3D_2(6)$; $C_3(8)$; $5C_2(12)$
209	$F432$	O^3	$2O(4)$; $T(8)$; $D_2(24)$; $C_4(24)$; $C_3(32)$; $3C_2(48)$ $[\div 4]$
210	$F4_132$	O^4	$2T(8)$; $2D_3(16)$; $C_3(32)$; $2C_2(48)$ $[\div 4]$
211	$I432$	O^5	$O(2)$; $D_4(6)$; $D_3(8)$; $D_2(12)$; $C_4(12)$; $C_3(16)$; $3C_2(24)$ $[\div 2]$
212	$P4_332$	O^6	$2D_3(4)$; $C_3(8)$; $C_2(12)$
213	$P4_132$	O^7	$2D_3(4)$; $C_3(8)$; $C_2(12)$
214	$I4_132$	O^8	$2D_3(8)$; $2D_2(12)$; $C_3(16)$; $3C_2(24)$ $[\div 2]$
215	$P\bar{4}3m$	T_d^1	$2T_d$; $2D_{2d}(3)$; $C_{3v}(4)$; $2C_{2v}(6)$; $C_2(12)$; $C_s(12)$
216	$F\bar{4}3m$	T_d^2	$4T_d(4)$; $C_{3v}(16)$; $2C_{2v}(24)$; $C_s(48)$ $[\div 4]$
217	$I\bar{4}3m$	T_d^3	$T_d(2)$; $D_{2d}(6)$; $C_{3v}(8)$; $S_4(12)$; $C_{2v}(12)$; $C_2(24)$; $C_s(24)$ $[\div 2]$
218	$P\bar{4}3m$	T_d^4	$T(2)$; $D_2(6)$; $2S_4(6)$; $C_3(8)$; $3C_2(12)$
219	$F\bar{4}3c$	T_d^5	$2T(8)$; $2S_4(24)$; $C_3(32)$; $2C_2(48)$ $[\div 4]$
220	$I\bar{4}3d$	T_d^6	$2S_4(12)$; $C_3(16)$; $C_2(24)$ $[\div 2]$
221	$Pm3m$	O_h^1	$2O_h$; $2D_{4h}(3)$; $2C_{4v}(6)$; $C_{3v}(8)$; $3C_{2v}(12)$; $3C_s(24)$
222	$Pn3n$	O_h^2	$O(2)$; $D_4(6)$; $C_{3i}(8)$; $S_4(12)$; $C_4(12)$; $C_3(16)$; $2C_2(24)$
223	$Pm3n$	O_h^3	$T_h(2)$; $D_{2h}(6)$; $2D_{2d}(6)$; $D_3(8)$; $3C_{2v}(12)$; $C_3(16)$; $C_2(24)$; $C_s(24)$
224	$Pn3m$	O_h^4	$T_d(2)$; $2D_{3d}(4)$; $D_{2d}(6)$; $C_{3v}(8)$; $D_2(12)$; $C_{2v}(12)$; $3C_2(24)$; $C_s(24)$
225	$Fm3m$	O_h^5	$2O_h(4)$; $T_d(8)$; $D_{2h}(24)$; $C_{4v}(24)$; $C_{3v}(32)$; $3C_{2v}(48)$; $2C_s(96)$ $[\div 4]$
226	$Fm3c$	O_h^6	$O(8)$; $T_h(8)$; $D_{2d}(24)$; $C_{4h}(24)$; $C_{2v}(48)$; $C_4(48)$; $C_3(64)$; $C_2(96)$; $C_s(96)$ $[\div 4]$
227	$Fd3m$	O_h^7	$2T_d(8)$; $2D_{3d}(16)$; $C_{3v}(32)$; $C_{2v}(48)$; $C_s(96)$; $C_2(96)$ $[\div 4]$
228	$Fd3c$	O_h^8	$T(16)$; $D_3(32)$; $C_{3i}(32)$; $S_4(48)$; $C_3(64)$; $2C_2(96)$ $[\div 4]$
229	$Im3m$	O_h^9	$O_h(2)$; $D_{4h}(6)$; $D_{3d}(8)$; $D_{2d}(12)$; $C_{4v}(12)$; $C_{3v}(16)$; $2C_{2v}(24)$; $C_2(48)$; $2C_s(48)$ $[\div 2]$
230	$Ia3d$	O_h^{10}	$C_{3i}(16)$; $D_3(16)$; $D_2(24)$; $S_4(24)$; $C_3(32)$; $2C_2(48)$ $[\div 2]$

References

1. R. C. C. Leite, R. S. Moore, S. P. S. Porto, and J. E. Ripper, *Phys. Rev. Letters*, **14,** 3 (1965)
2. E. B. Wilson, J. C. Decius, and P. C. Cross, *Molecular Vibrations*, McGraw-Hill, New York (1955)
3. G. Placzek, *Marx Handbuch der Radiologie*, **6** (1934) [Also available as a translation: *Rayleigh and Raman Scattering*, U.S. At. Energy Comm., UCRL-TRANS-526 (L) (1962)]
4. H. Szymanski (Ed.), *Raman Spectroscopy*, Plenum Press, New York (1967)
5. P. P. Shorygin and T. M. Ivanovna, *Optics and Spectroscopy*, **25,** 200 (1968)
6. G. Herzberg, *Infrared and Raman Spectra of Polyatomic Molecules*, Van Nostrand, New York (1945)
7. D. Ross, *Siemens Review*, **34,** 159 (1967), and refs. therein
8. M. V. Evans, T. M. Hard, and W. F. Murphy, *J. Opt. Soc. Amer.*, **56,** 1638 (1966)
9. R. C. Hawes, K. P. George, D. C. Nelson, and R. Beckwith, *Analyt. Chem.*, **38,** 1842 (1966)
10. H. Cary, W. S. Galloway, and K. P. George, Special Publ. Cary Instrument Co., Monrovia, California, 1962
11. M. Delhaye and M. Migeon, *Compt. Rend.*, **262,** 1513 (1966)
12. D. Landon and S. Porto, *Appl. Optics*, **4,** 762 (1965)
13. Jarrell-Ash Co., *Spectrum Scanner*, **22,** No. 3, 5 (1967)
14. A. Weber, *Spex Speaker*, **9,** No. 4 (1966)
15. W. G. Fastie, *J. Opt. Soc. Amer.*, **42,** 641, 647 (1952)
16. G. A. Ozin, private communication
17. H. J. Smith and J. P. Rodman, *Appl. Optics*, **2,** 181 (1963)
18. J. K. Nakamura and S. E. Schwarz, *Appl. Optics*, **7,** 1073 (1968)
19. Yo-Han Pao, R. N. Zitter, and J. E. Griffiths, *J. Opt. Soc. Amer.*, **56,** 1133 (1966)
20. S. P. S. Porto and D. L. Wood, *J. Opt. Soc. Amer.*, **52,** 251 (1962)
21. B. Stoicheff, 10th Int. Spectr. Colloquium, Univ. of Md., June 1962, Spartan Books, Washington, D.C. (1963)
22. G. E. Daniliheva, B. A. Zubov, M. M. Sushchinskij, and I. K. Shuvalov, *Soviet Phys. JETP*, **17,** 1473 (1963)
23. H. Kogelnick and S. P. S. Porto, *J. Opt. Soc. Amer.*, **53,** 1446 (1963)
24. S. P. S. Porto, L. E. Cheeseman, and J. B. Siqueira, Symp. Mol. Struct. and Spectr., Ohio State Univ., June 1963
25. S. P. S. Porto, *Ann. N.Y. Acad. Sci.*, **122,** 613 (1965)
26. R. C. Leite and S. P. S. Porto, *J. Opt. Soc. Amer.*, **54,** 981 (1964)
27. R. C. Leite and S. P. S. Porto, *J. Opt. Soc. Amer.*, **53,** 1503 (1963)

28. J. A. Koningstein and R. Smith, *J. Opt. Soc. Amer.*, **54,** 569 (1964)
29. H. Siegler, C. D. Hinman, and A. F. Slomba, Symp. Mol. Struct and Spectr., Ohio State Univ., June 1964; Opt. Soc. Amer. Conference, Washington, D.C., April 1964; *Perkin-Elmer Instr. News*, **15,** No. 2, 6; No. 3, 1 (1964); **16,** No. 2, 16 (1965)
30. G. W. Chantry, H. A. Gebbie, and C. Hilsun, *Nature*, **203,** 1052 (1964)
31. A. Weber and S. P. S. Porto, *J. Opt. Soc. Amer.*, **55,** 1033 (1965)
32. B. Schrader and M. Stockburger, *Z. Analyt. Chem.*, **216,** 117 (1966)
33. B. Schrader and M. Stockburger, *Naturwiss.*, **52,** 298 (1965)
34. M. Delhaye and M. Migeon, *Compt. Rend.*, (*a*) **261,** 2613 (1965); (*b*) **262,** 702 (1966)
35. A. Lau and J. H. Hertz, *Spectrochim. Acta*, **22,** 1935 (1966); *Monatsh.*, **8,** 762 (1966)
36. J. Brandmüller, K. Burchardi, H. Hacker, and H. W. Schrotter, *Z. Angew. Phys.*, **23,** 112 (1967)
37. M. Delhaye and M. Bridoux, Colloqu. Raman Spectr., Magdeburg, October 1966
38. R. C. C. Leite, R. S. Moore, and S. P. S. Porto, *J. Chem. Phys.*, **40,** 3741 (1964)
39. T. C. Damen, R. C. C. Leite, and S. P. S. Porto, *Phys. Rev. Letters*, **14,** 9 (1965)
40. I. R. Beattie, *Chem. in Britain*, **3,** 347 (1967)
41. J. Bryant, *Spectrochim. Acta*, **24***A*, 9 (1968)
42. H. Clase, Dept. of Chemistry, University of Sussex, Brighton, private communication
43. E. J. Loader, *Handbook of Raman Spectroscopy*, Heyden, London (1969)
44. G. F. Bailey, S. Kint, and J. R. Scherer, *Analyt. Chem.*, **39,** 1040 (1967)
45. P. J. Hendra and E. J. Loader, *Chem. and Ind.*, 719 (1968)
46. M. Qurashi, Ph.D. Thesis, Southampton, (1970)
47. S. K. Freeman and D. O. Landon, *Spex Speaker*, **8,** 4 (1968)
48. H. Haber and H. Sloane, Special Publ. Cary Instrument Co., R-68-2 (1968)
49. E. R. Lippincott and R. D. Fisher, *Spectrochim. Acta*, **6,** 255 (1954)
50. M. J. Gall, private communication
51. M. Delhaye, Inst. Petroleum Mol. Spectr. Congress, Brighton, 1968 [*Proc. Inst. Pet.* (1968)]
52. M. Bridoux, *Rev. Opt.*, **8,** 389 (1967)
53. I. R. Beattie and M. J. Gall, *Inorg. Nucl. Chem. Letters*, **4,** 677 (1968)
54. L. A. Woodward and D. A. Long, *Trans. Faraday Soc.*, **45,** 1131 (1949)
55. S. P. S. Porto, *J. Opt. Soc. Amer.*, **56,** 1585 (1966)
56. I. I. Kondilenko and P. A. Korotkov, *Optics and Spectroscopy*, **24,** 246 (1968)
57. I. R. Beattie, T. Gilson, K. Livingston, V. Fawcett, and G. Ozin, *J. Chem. Soc.*,*A*, 712 (1967)
58. J. T. Edsall and E. B. Wilson, *J. Chem. Phys.*, **6,** 124 (1938)
59. H. J. Bernstein and G. Allen, *J. Opt. Soc. Amer.*, **45,** 237 (1955)
60. H. W. Schrötter and H. J. Bernstein, *J. Mol. Spectr.*, **12,** 1 (1964)
61. J. Brandmüller and H. Moser, *Einführung in die Raman Spektroskopie*, Steinkopff, Darmstadt (1962)
62. D. G. Rea, *J. Opt. Soc. Amer.*, **49,** 90 (1959)

63. T. T. Wall and D. F. Hornig, *J. Chem. Phys.*, **45,** 3424 (1966)
64. W. F. Murphy, M. V. Evans, and P. Bender, *J. Chem. Phys.*, **47,** 1836 (1967)
65. J. R. Nielsen, *J. Opt. Soc. Amer.*, **20,** 701 (1930)
66. I. A. Stenhouse, Ph.D. Thesis, Cambridge (1968)
67. L. Baxter, *J. Opt. Soc. Amer.*, **46,** 435 (1956)
68. I. R. Beattie, M. J. Gall, and G. A. Ozin, *J. Chem. Soc., A*, 1001 (1969)
69. P. J. Hendra and M. Qurashi, *J. Chem. Soc.,A*, 2963 (1968)
70. D. D. Tunnicliff and A. C. Jones, *Spectrochim. Acta*, **18,** 579 (1962)
71. D. A. Bahnick and W. B. Person, *J. Chem. Phys.*, **48,** 1251 (1968)
72. C.-H. Ting, *Spectrochim. Acta*, **24*A*,** 1177 (1968)
73. A. C. Albrecht, *J. Chem. Phys.*, **34,** 1496 (1961)
74. L. L. Krushinskii and P. P. Shorygin, *Tr. Komis. po Spektroskopii, Akad. Nauk SSSR*, (1) 162 (1964) [*Chem. Abs.*, **63,** 7792 (1965)]
75. K. Rebane and T. Rebane, *Eesti NSV Teaduste Akad. Toimetised, Tehniliste Fuusikalis-Mat. Teaduste Seeria*, **12,** 227 (1963) [*Chem. Abs.*, **60,** 7589 (1964)]
76. G. Eckardt and W. G. Wagner, *J. Mol. Spectr.*, **19,** 407 (1966)
77. T. P. Tulub and Ya. S. Bobovich, *Optics and Spectroscopy*, Suppl. 2, 161 (1966)
78. G. Fini, P. Mirone, and P. Patella, *J. Mol. Spectr.*, **28,** 144 (1968)
79. M. Eliashevich and M. Wolkenstein, *J. Phys. USSR*, **9,** 101, 326 (1945)
80. L. M. Sverdlov and N. I. Prokofeva, *Optics and Spectroscopy*, **18,** 18 (1965)
81. L. M. Sverdlov, *Optics and Spectroscopy*, **15,** 439 (1963)
82. R. N. Jones and M. K. Jones, *Analyt. Chem.*, **38,** 393 R (1966)
83. M. S. Child and H. C. Longuet-Higgins, *Phil. Trans. Roy. Soc.,A*, **254,** 259 (1961)
84. R. G. Gordon, *J. Chem. Phys.*, **43,** 1307 (1965)
85. P. W. Higgs, *Proc. Roy. Soc.,A*, **220,** 472 (1953)
86. R. Loudon, *Adv. Phys.*, **13,** 423 (1964); errata, *Adv. Phys.*, **14,** 621 (1965)
87. S. Bhagavantam, *Crystal Symmetry and Physical Properties*, Academic Press, London (1966)
88. O. Brafman and S. S. Mitra, *Phys. Rev.*, **171,** 931 (1968)
89. C. H. Henry and J. J. Hopfield, *Phys. Rev. Letters*, **15,** 964 (1965)
90. S. P. S. Porto, B. Tell, and T. C. Damen, *Phys. Rev. Letters*, **16,** 450 (1966)
91. R. P. Bauman and S. P. S. Porto, *Phys. Rev.*, **161,** 842 (1967)
92. R. K. Chang, B. Lacina, and P. S. Pershan, *Phys. Rev. Letters*, **17,** 755 (1966)
93. O. Brafman, I. F. Chang, G. Lengyel, S. S. Mitra, and E. Carnall, *Phys. Rev. Letters*, **19,** 1120 (1967)
94. J. F. Parrish and C. H. Perry, private communication
95. R. Braunstein, *Phys. Rev.*, **130,** 879 (1963)
96. J. M. Worlock and S. P. S. Porto, *Phys. Rev. Letters*, **15,** 697 (1965)
97. O. Brafman and S. S. Mitra, Proc. Int. Conf. on Light Scattering Spectra of Solids, New York, September 1968
98. I. R. Beattie and D. J. Reynolds, *Chem. Comm.*, 1531 (1968)
99. L. Couture–Mathieu and J.-P. Mathieu, *Acta Cryst.*, **5,** 571 (1952)
100. R. S. Halford, *J. Chem. Phys.*, **14,** 8 (1946)
101. I. R. Beattie, T. Gilson, and G. A. Ozin, *J. Chem. Soc.,A*, 534 (1969)

102. S. S. Mitra, *Solid State Physics*, **13,** 1—80 (1962); S. S. Mitra and P. J. Gielisse, *Progress in Infrared Spectroscopy* (ed. H. Szymanski), Plenum Press, New York (1963)
103. I. R. Beattie and T. Gilson, *Proc. Roy. Soc.,A*, **307,** 407 (1968)
104. I. R. Beattie, T. Gilson, and G. A. Ozin, *J. Chem. Soc.,A*, 2765 (1968)
105. I. R. Beattie and T. Gilson, *J. Chem. Soc.,A*, 2322 (1969)
106. G. W. Cohen–Solal, *Compt. Rend.*, **264,** 537 (1967)
107. E. J. Ambrose, A. Elliott, and R. B. Temple, *Proc. Roy. Soc.,A*, **206,** 192 (1951)
108. G. C. Pimentel, *J. Chem. Phys.*, **19,** 1536 (1951)
109. I. R. Beattie and G. A. Ozin, *J. Chem. Soc.,A*, 542, (1969)
110. M. Suzuki, T. Yokayama, and M. Ito, *Spectrochim. Acta*, **24*A*,** 1091 (1968)
111. G. A. Ozin, *J. Chem. Soc.,A*, 116 (1969)
112. T. C. Damen, S. P. S. Porto, and B. Tell, *Phys. Rev.*, **142,** 570 (1966)
113. S. P. S. Porto, J. A. Giordmaine, and T. C. Damen, *Phys. Rev.*, **147,** 608 (1966)
114. V. Chandrasekharan, *Z. Phys.*, **154,** 43 (1959), and refs. therein
115. H. Poulet and J.-P. Mathieu, *J. Chim. Phys.*, **60,** 442 (1963)
116. B. P. Stoicheff, *Adv. Spectr.*, 1 (1964)
117. S. P. S. Porto and R. S. Krishnan, *J. Chem. Phys.*, **47,** 1009 (1967)
118. R. F. Schaufele and M. J. Weber, *Phys. Rev.*, **152,** 705 (1966)
119. R. F. Schaufele and M. J. Weber, *J. Chem. Phys.*, **46,** 2859 (1967)
120. J. P. Hurrell, S. P. S. Porto, I. F. Chang, S. S. Mitra, and R. P. Bauman, *Phys. Rev.*, **173,** 851 (1968)
121. O. Brafman, G. Lengyel, S. S. Mitra, P. J. Gielisse, J. N. Plendl, and L. C. Mansur, *Solid State Comm.*, **6,** 523 (1968)
122. A. R. Gee and G. W. Robinson, *J. Chem. Phys.*, **46,** 4847 (1967)
123. S. P. S. Porto, P. A. Fleury, and T. C. Damen, *Phys. Rev.*, **154,** 522 (1967)
124. S. M. Shapiro, D. C. O'Shea, and H. Z. Cummins, *Phys. Rev. Letters*, **19,** 361 (1967)
125. S. P. S. Porto and J. F. Scott, *Phys. Rev.*, **157,** 716 (1967)
126. P. A. Fleury, S. P. S. Porto, L. E. Cheeseman, and H. J. Guggenheim, *Phys. Rev. Letters*, **17,** 84 (1966)
127. P. A. Fleury, S. P. S. Porto, and R. Loudon, *Phys. Rev. Letters*, **18,** 658 (1967)
128. R. F. Schaufele, M. J. Weber, and B. D. Silverman, *Phys. Letters*, **25*A*,** 47 (1967)
129. W. G. Nilsen and J. G. Skinner, *J. Chem. Phys.*, **47,** 1413 (1967)
130. S. S. Mitra, *J. Phys. Soc. Japan* (*Suppl.*), **21,** 61 (1966)
131. J. L. Birman and A. K. Ganguly, *Phys. Rev. Letters*, **17,** 647 (1966)
132. B. P. Stoicheff, *Adv. Spectr.*, 91 (1959)
133. B. P. Stoicheff, in *Experimental Physics: Molecular Physics* (ed. D. Williams), Academic Press, New York, **3,** 111 (1962)
134. S. P. S. Porto, L. E. Cheeseman, A. Weber, and J. J. Barrett, *J. Opt. Soc. Amer.*, **57,** 19 (1967)
135. J. J. Barrett and N. I. Adams III, *J. Opt. Soc. Amer.*, **58,** 311 (1968)
136. A. Weber and S. P. S. Porto, 8th Europ. Congr. Mol. Spectr., Copenhagen, 1965, Paper 415
137. A. Weber and S. P. S. Porto, *Bull. Amer. Phys. Soc.*, **10,** 101 (1965)

138. G. Herzberg, *Spectra of Diatomic Molecules*, Van Nostrand, New York (1950): (*a*) pp. 114, 121; (*b*) p. 130; (*c*) p. 133
139. H. J. Bernstein, 9th Europ. Congr. Mol. Spectr., Madrid, 1967, Paper 263
140. J. J. Barrett and J. D. Rigden, 9th Europ. Congr. Mol. Spectr., Madrid, 1967, Paper 264
141. E. L. Gasner and H. H. Claassen, *Inorg. Chem.*, **6,** 1937 (1967)
142. L. S. Bantell, *J. Chem. Phys.*, **46,** 4530 (1967)
143. Jarrell-Ash Co., Tech. Bull. EB-140 on the 25-300 instrument
144. D. A. Leonard, *Nature*, **216,** 142 (1967)
145. M. Delhaye, *Appl. Optics*, **7,** 2195 (1968)
146. J. I. Steinfeld, R. N. Zane, L. Jones, M. Lesk, and W. Klemperer, *J. Chem. Phys.*, **42,** 25 (1965)
147. I. R. Beattie and J. Horder, *J. Chem. Soc.,A*, 2655 (1969)
148. A. Muller and M. Stockburger, *Z. Naturforsch.*, **20**a, 1242 (1965)
149. B. Schrader and M. Stockburger, *Naturwiss.*, **52,** 298 (1965); *Z. Analyt. Chem.*, **216,** 117 (1966)
150. P. J. Hendra, *Nature*, **212,** 179 (1966)
151. P. J. Hendra, *J. Chem. Soc.,A*, 1298 (1967)
152. P. J. Hendra and Z. Jovic, *J. Chem. Soc.,A*, 735 (1967); 911 (1968)
153. P. Klaboe, *J. Amer. Chem. Soc.*, **89,** 3667 (1967)
154. P. J. Hendra, *Spectrochim., Acta*, **24***A*, 125 (1968)
155. E. A. V. Ebsworth, in *Infrared Spectroscopy and Molecular Structure* (ed. M. Davies), Elsevier, Amsterdam (1963)
156. N. Bacon, A. J. Boulton, and A. R. Katritzky, *Trans. Faraday Soc.*, **63,** 833 (1967)
157. G. C. Hayward and P. J. Hendra, *J. Chem. Soc.,A*, 1760 (1969)
158. G. C. Hayward and P. J. Hendra, *Spectrochim., Acta*, **23***A*, 2309 (1967)
159. J. C. Evans and G. Y.-S. Lo, *Inorg. Chem.*, **6,** 1483 (1967)
160. J. R. Allkins, F. J. Blunt, P. J. Hendra, and P. J. Park, *J. Mol. Struct.*, in the press
161. E. W. Abel, R. A. N. McLean, S. P. Tyfield, P. S. Braterman, A. P. Walker, and P. J. Hendra, *J. Mol. Spectr.*, **30,** 29 (1969)
162. P. Klaboe and E. Klöster-Jensen, *Spectrochim. Acta*, **23***A*, 1981 (1967)
163. P. J. Hendra and J. R. Mackenzie, *Chem. Comm.*, 760 (1968)
164. K. W. F. Kohlrausch, *Ramanspektren*, Akad Verlag Becker and Erler, Leipzig (1943)
165. K. Nakamoto, *Infrared Spectra of Inorganic and Coordination Compounds*, Wiley, New York (1963)
166. P. J. Hendra and P. J. Park, *Spectrochim. Acta*, **25***A*, 227 (1969)
167. P. J. Hendra and P. J. Park, *Spectrochim. Acta*, **25***A*, 909 (1969)
168. I. R. Beattie and co-workers, *J. Chem. Soc.,A*, 712 (1967); 1092, 2373, 2772 (1968); 482 (1969)
169. P. J. Hendra and P. J. Park, *J. Chem. Soc.,A*, 908 (1968)
170. E. B. Bradley, M. S. Mathur, and C. A. Frenzel, *J. Chem. Phys.*, **47,** 4325 (1967)
171. R. F. Schaufele and T. Shimanouchi, *J. Chem. Phys.*, **47,** 3605 (1967)
172. R. F. Schaufele, *J. Chem. Phys.*, **49,** 4168 (1968)
173. C. Troyanowsky, 9th Europ. Congr. Mol. Spectr., Madrid, 1967, Paper 265

174. M. Tobin, *Science*, **161,** 68 (1968)
175. R. C. Lord and G. J. Thomas, Jr., *Spectrochim. Acta*, **23A**, 2551 (1967); *Biochim. Biophys. Acta*, **142,** 1 (1967)
176. G. E. Walrafen, *J. Chem. Phys.*, **46,** 1870 (1967)
177. G. E. Walrafen, *Hydrogen Bonded Systems*, Taylor and Francis, London, 1968, p. 9
178. R. E. Hester and V. E. L. Grossman, *Spectrochim. Acta*, **23A**, 1945 (1967)
179. G. J. Janz and S. C. Wait, Jr., *Raman Spectroscopy* (ed. H. Szymanski), Plenum Press, New York (1967)
180. J. H. R. Clarke, C. Solomons, and K. Balasubrahmanyan, *Rev. Sci. Instr.*, **38,** 655 (1967); *J. Chem. Phys.*, **47**, 1823 (1967)
181. J. H. R. Clarke and R. E. Hester, *Chem. Comm.*, 1042 (1968); *J. Chem. Phys.*, **50** (1969)
182. J. H. R. Clarke, C. Solomons, and J. O'M. Bockris, *J. Chem. Phys.*, **49,** 445 (1968)
183. A. J. Melveger, R. K. Khanna, B. R. Guscott, and E. R. Lippincott, *Inorg. Chem.*, **21,** 2145 (1968)
184. D. A. Jennings, M. McClintock, and M. Mizushima, 9th Europ. Congr. Mol. Spectr., Madrid, 1967, Paper 262
185. C. Postmus, V. A. Maroni, J. R. Ferraro, and S. S. Mitra, *Inorg. Nucl. Chem. Letters*, **4,** 269 (1968)
186. E. R. Lippincott and co-workers, *Inorg. Chem.*, **7**, 1630 (1968); *Chem. Phys. Letters*, **2**, 99 (1968)
187. A. G. Maki and R. Forneris, *Spectrochim. Acta*, **23A,** 867 (1967)
188. R. F. Schaufele, *Trans. N.Y. Acad. Sci.*, **30** (1967)
189. P. J. Hendra, *Adv. Polymer Sci.*, **6,** 151 (1969)
190. R. Signer and J. Weiler, *Helv. Chim. Acta*, **15**, 649 (1932)
191. L. I. Maklakov and V. N. Nikitin, *Optics and Spectroscopy*, **17**, 242 (1964)
192. J. R. Nielsen, *J. Polymer Sci.,C*, 19 (1963)
193. R. F. Schaufele, *J. Opt. Soc. Amer.*, **57,** 105 (1967)
194. P. J. Hendra and H. A. Willis, *Chem. and Ind.*, 2146 (1967)
195. P. J. Hendra and H. A. Willis, *Chem. Comm.*, 225 (1968)
196. P. J. Hendra, Rep. Conf. Inst. Pet., Brighton, 1968
197. T. Shimanouchi, *J. Polymer Sci.,C*, **85** (1963)
198. R. Sneider, *J. Mol. Spectr.*, **23,** 224 (1967)
199. P. J. Hendra, *J. Mol. Spectr.*, **28,** 118 (1968)
200. S. Krimm and C. G. Opasker, *Spectrochim. Acta*, **21,** 1165 (1965)
201. T. P. Linn and J. L. Koenig, *J. Mol. Spectr.*, **9,** 228 (1962)
202. J. H. Schachtsneider and R. G. Sneider, *Spectrochim. Acta*, **19,** 117 (1963)
203. C. A. Frenzel, E. B. Bradley, and M. S. Mathur, *J. Chem. Phys.*, **49,** 3789 (1968)
204. G. Zerbi and P. J. Hendra, *J. Mol. Spectr.*, **27,** 17 (1968)
205. J. H. Schachtsneider and R. G. Sneider, *Spectrochim. Acta*, **21,** 1527 (1965)
206. M. Peraldo and M. Cambini, *Spectrochim. Acta*, **21,** 1509 (1965)
207. J. L. Koenig and S. W. Cornell, *J. Appl. Phys.*, **39,** 4883 (1968)
208. F. J. Boerio and J. L. Koenig, *J. Chem. Phys.*, in press (1970)
209. P. J. Hendra, H. A. Willis, and H. Zichi, *Polymer*, in the press
210. N. Miyazawa, T. Shimanouchi, and S. Mizushima, *J. Chem. Phys.*, **24,** 408 (1956)

211. J. L. Koenig and F. J. Boerio, *J. Chem. Phys.*, **50,** 2823 (1969); M. J. Hannon, F. J. Boerio, and J. L. Koenig, *J. Chem. Phys.*, **50**, 2829 (1969); P. J. Hendra, C. J. Peacock, M. J. Cadby, and H. A. Willis, *J. Chem. Soc.*, in press (1970)
212. P. J. Hendra, P. Holliday, and J. R. Mackenzie, *Spectrochim. Acta*, **25*A*,** 1349 (1969)
213. L. H. Little, *Infrared Spectra of Adsorbed Species*, Academic Press, London (1968)
214. M. L. Hair, *Infrared Spectroscopy in Surface Chemistry*, Marcel Dekker, New York (1967)
215. G. Karagounis and R. Issa, *Nature*, **195,** 1196 (1962)
216. G. Karagounis and R. Issa, *Z. Elektrochem.*, **66,** 874 (1962)
217. P. J. Hendra and E. J. Loader, *Nature*, **216,** 789 (1967)
218. P. J. Hendra and E. J. Loader, *Nature*, **217,** 637 (1968)
219. P. J. Hendra and E. J. Loader, *Trans. Faraday Soc.*, in the press
220. E. V. Pershina and Sh. Sh. Raskin, *Optics and Spectroscopy*, Suppl. 3; *Mol. Spectr.*, **2,** 328; *Trans.*, 114 (1967)
221. J. A. Koningstein, *J. Opt. Soc. Amer.*, **56,** 1402 (1966)
222. J. A. Koningstein, *J. Chem. Phys.*, **46,** 2811 (1967)
223. C. Brecker, H. Samelson, A. Lempicki, R. Riley, and T. Peters, *Phys. Rev.*, **155,** 178 (1967)
224. J. A. Koningstein, private communication
225. O. S. Mortensen and J. A. Koningstein, *Chem. Phys. Letters*, **1,** 409 (1967)
226. J. A. Koningstein and O. S. Mortensen, *Phys. Rev.*, **168,** 75 (1968)
227. J. A. Koningstein and O. S. Mortensen, *Chem. Phys. Letters*, **1,** 693 (1968)
228. E. J. Woodbury and W. K. Ng, *Proc. Inst. Rad. Eng.*, **50,** 2367 (1962)
229. H. W. Schrotter, *Naturwiss.*, **23,** 607 (1967)
230. B. P. Stoicheff, *Phys. Letters*, **7,** 186 (1963)
231. R. Y. Chiao, M. A. Johnson, S. Krinsky, H. A. Smith, C. H. Townes, and E. Garmire, *IEEE J. Quant. Electronics*, **2,** 267 (1966), and refs. therein
232. G. G. Bret and M. M. Denariez, *Appl. Phys. Letters*, **8,** 151 (1966)
233. G. G. Bret and M. M. Denariez, *Phys. Letters*, **42,** 583 (1966)
234. M. Maier, *Phys. Letters*, **20,** 388 (1966)
235. R. W. Hellworth, *Phys. Rev.*, **130,** 1850 (1963)
236. G. Bret and G. Mayer, in *Physics of Quantum Electronics* (ed. P. L. Kelly), McGraw-Hill, New York (1966), p. 180
237. G. Bret, *Compt. Rend.*, **260,** 6323 (1965)
238. R. W. Hellworth, F. J. McClung, W. G. Wagner, and D. Weiner, *Bull Amer. Phys. Soc.*, **9,** 490 (1964)
239. E. Garmire, *Phys. Letters*, **17,** 251 (1965)
240. E. R. Lippincott, G. E. Meyers, and P. J. Hendra, *Spectrochim. Acta*, **22,** 1493 (1966)
241. R. W. Minck, E. E. Hagenlocker, and W. G. Rado, *Phys. Rev. Letters*, **17,** 229 (1966)
242. P. P. Sorokin et al., *Appl. Phys. Letters*, **10,** 44 (1967)
243. M. Rokni and S. Yatsiv, *Phys. Letters*, **24*A*,** 277 (1967)
244. M. Rokni and S. Yatsiv, *IEEE J. Quantum Electronics*, **3,** 329 (1967)
245. M. D. Martin and E. L. Thomas, *Phys. Letters*, **5,** 58 (1964)
246. J. A. Duardo and F. M. Johnson, *J. Chem. Phys.*, **45,** 2325 (1965)

247. S. Yoshikawa, Y. Matsumura, and H. Inaba, *Appl. Phys. Letters*, **8**, 27 (1966)
248. M. Maier, W. Kaiser, and J. A. Giordmaine, *Phys. Rev. Letters*, **17**, 1275 (1966)
249. F. De Martini and J. Ducuing, *Phys. Rev. Letters*, **17**, 117 (1966)
250. W. A. Zubov, G. W. Peregudow, M. M. Sushchinskij, W. A. Tzchirkow, and I. K. Shuvalov, *Pismav Redakjiju ZETF*, **5**, 188 (1967)
251. Li Yin-Yuan, *Acta Phys. Sinica*, **20**, 164 (1964)
252. S. Kielich, *Bull. Acad. Pol. Sci.*, **12**, 53 (1964)
253. S. J. Cyvin, J. E. Rauch, and J. C. Decius, *J. Chem. Phys.*, **43**, 4083 (1964)
254. R. W. Terhune, P. D. Maker, and C. M. Savage, *Phys. Rev. Letters*, **14**, 681 (1965)
255. P. D. Maker, in *Physics of Quantum Electronics* (see ref. 236), p. 60
256. W. J. Jones and B. P. Stoicheff, *Phys. Rev. Letters*, **13**, 657 (1964)
257. A. K. Macquillan and B. P. Stoicheff, in *Physics of Quantum Electronics* (see ref. 236), p. 192
258. S. Dumartin, B. Oksengorn, and B. Vodar, *Compt. Rend.*, **261**, 3767 (1965)
259. A. D. Buckingham, *J. Chem. Phys.*, **43**, 25 (1965)
260. L. S. Bartell and H. B. Thompson, in *Physics of Quantum Electronics* (see ref. 236), p. 192
261. J. R. Allkins and P. J. Hendra, *J. Chem. Soc.*, *A*, 1325 (1967)
262. M. Delhaye, N. Durrieu-Mercier, and M. Migeon, *Compt. Rend.*, *B*, **267**, 135 (1968)
263. J. C. Carter, R. F. Bratton, and J. R. Jackowitz, *J. Chem. Phys.*, **49**, 3752 (1968)
264. P. J. Hendra, J. R. Horder, and E. J. Loader, *Chem. Comm.*, in press (1970)
265. E. J. Loader, Ph.D. Thesis, Southampton (1970)
266. J. A. Koningstein and O. S. Mortensen, *Nature*, **217**, 445 (1968)
267. G. Eckhardt, R. W. Hellworth, F. J. McClung, S. E. Schwarz, D. Wiener, and E. J. Woodbury, *Phys. Rev. Letters*, **9**, 455 (1962)
268. R. W. Minck, R. W. Terhune, and W. G. Rado, *Appl. Phys. Letters*, **3**, 181 (1963)
269. M. Geller, D. P. Bortfield, and W. R. Sooy, *Appl. Phys. Letters*, **3**, 36 (1963)
270. H. W. Schrotter, *Naturwiss.*, **54**, 513 (1967)
271. G. Revoire, *J. Phys.* (Paris), **26**, 711 (1967)
272. R. G. Sneider, *J. Mol. Spectr.*, **31**, 464 (1969)

Subject Index

Accordion mode, 102, 161
Acoustic branch, 97, 112
Acoustic modes, 94, 97, 102, 103, 118
Adsorbed species, Raman of, 183
Al_2Br_6, 129, 137
Allowed combinations, 86–87, 95, 107, 116–119
Amplifier systems, 34
Analysis, qualitative, 165
Anatase, 112–114
Anharmonicity, 12, 17
Anisotropic scattering, 63, 70
Anisotropy of a tensor, 62, 69
Anthracene, 129
Anti-Stokes, 2, 10, 12, 17, 79
Aqueous systems, Raman of, 163
Average scattering tensor, 62

BaF_2, 105
Band contours, 83
  in liquid phase, 164
'Beat' frequencies, 9
Biological specimens, Raman of, 161, 162
Birefringence, 6, 78, 131, 134–138
Boltzmann distribution, 2, 12, 17
Born–Oppenheimer approximation, 80
Brewster angle, 6
Brillouin scattering, 94
Brillouin zone, 93, 94, 105, 112, 125
Bulk properties, 89, 92
Bromo-iodide ion, Raman spectrum, 156

CaF_2–SrF_2, 105
Calibration (frequency), 51
Calibration (intensity), 77–78
CdS–CdSe, 105
Charge-transfer complexes, Raman spectra and structures, 156
Chemisorption, 188
Cinnabar, 136
Classical treatment, 81
Coherence, 4–6, 15
Coloured compounds, Raman of, 153–155
Combination bands, 16, 17, 18, 126
Components (of degeneracy), 117
Compton effect, 197
Convergence errors, 63
Correlation splitting, 107, 117
  in polyethylene, 171
Correlation theorem, 86, 108, 116–125
Cosecant bar wavenumber drive system, facing p. 50
Coupling, 94, 103, 116–119, 126
Crystal axes, 84, 109–110, 115, 128, 134, 136
Crystal field, 84, 126–131
Crystal modes (vibrations), 84, 85, 93, 94–105, 116–119
  internal, 106–108, 116–119, 126–131
  external, 106–108, 116–119, 207–209
Crystal point symmetry, 84, 90
Crystal setting, 133–134
Crystal size, 84, 104, 133
Crystal symmetry, 85, 88–91, 98, 99, 100, 111
$CuCl_4^{2-}$, 109, 119, 120, 121
$CuCl_2,2H_2O$, 119
Cyclic approximation, 89, 92
Cyclic boundary conditions, 92

D.c. amplification, 34
Degrees of freedom, 92, 93, 94, 112, 118
Depolarization ratios, 61, 65–67, 69, 130–131, 139
  of gases, 144, 146

Detectors: Photomultipliers, 32
  cooling of, 33
  Image intensifiers, 59
  Vidicon TV camera tubes, 59
'Dichroism' in Raman spectra of oriented specimens, 169
Difference bands, 16, 19
Disorder, 104
Dispersion, 95, 96, 102, 105
Divergence errors, 64, 70–78, 132–133

Electric vector, 4–6
Electromagnetic radiation, 3–8
Electronic transitions, Raman of, 189
Electrostatic forces, long-range, 98–100
End effects, 100–101
Enzymes, Raman of, 162
Ewald–Oseen theory, 14
Explosives, Raman of, 157, 158
External modes, 106–108, 116–119, 207–209
External resonator arrangement, 71
External standards, 77

Factor group, 89–91, 93, 94, 95, 99, 101, 103, 104, 108
  analysis, 108–125, 126
  fundamentals, 93, 94, 96, 99, 112
  in polyethylene, 172
F-centres, 105
Fibres, Raman of, 171
Finite group, 86, 89
Finite space, 89, 92
Fluids, 62
Fluorescence, 1, 2
  resonance, 151
Fluorescent compounds, 156, 157
Franck–Condon integrals, 80

Ga_2Cl_6, 129
Garnets, 189, 190, 191
Gas-phase Raman, experimental methods, 143–145
  vibrational spectra, 147, 148, 151
Geometrical-optical effects, 70–78
Glasses, 78
Glide planes, 90
Group, 86

Half-wave plate, 6–8, 67
'Handed' vibrations, 119
Harmonic generations in lasers, 25, 194
Hexanitrosobenzene, structure, 155
Hg_2Cl_2, 115, 119
Highly reactive compounds, Raman of, 159
High-pressure experiments, 165
High-speed Raman spectrometry, 58–60
High-temperature species, 163
  melts, 164
Hindered rotation, 116
'Hot' bands, 12, 16, 19
Hybrid modes, 98–100
Hyperpolarizability, 15
Hyper-Raman effect, 16, 18, 127, 194, 196

Identity operation, 86
Improper rotation, 111
$InCl_3$,$2NMe_3$, 129
Indicatrix (optical), 134–136
Inelastic neutron scattering, 93
Infinite crystal approximation, 89
Infinite group, 86, 89, 115
Infrared activity, 97–100, 126–131, 207–209
Intensity (definition), 3
Interferometers, in Raman spectroscopy, 36
  gas-phase applications, 152
Internal field correction, 77, 81
Internal modes, 106–108, 116–119, 126–131
Internal standards, 77
Invariants (of a tensor), 62
Inverse Raman effect, 16, 195
Irradiance, 3
Irreducible representation, 19, 85, 86, 90, 96, 112, 116–119
Isotropic scattering, 63, 70

Laboratory axes, 63
Lasers, general description, 201–206
  argon ion, 24, 143, 204
  helium–neon, 24, 25, 143, 203
  krypton ion, 24, 204

Lasers (*cont.*)
  neodymium, 25, 203
  nitrogen, 148
  Q-spoiled, 25
  ruby, 24, 201
  tunable, 25
Laser Raman spectrometers, general performance, 49–58
  historical development, 35
  Cary, 37
  Coderg, 39
  Huet, 42
  Japanese Optical Co., 44
  Jarrell-Ash, 42
  Perkin-Elmer, 45, 46
  Spectra Physics, 47
  Spex, 48
'Lattice' modes, 107
$LiClO_4,3H_2O$, 139
Lifetime of excited state, 1
Line group, 91, 107
  vibrations, 100–101
Linear groupings, 115
Local field correction, 77, 81
Local mode, 105
Longitudinal mode, 97–100
Low-temperature species, 163

Magnons, 140
Maxwell equations, 3, 15, 81
Mean value, 62
Melts, Raman of, 164
Mixed crystals, 105
Molecular crystals, 84
Monochromators, Czerny–Turner, 31
  discrimination against stray light, 30
  grating, 31
  prism, 22
Monoclinic crystals, 136–137
MoO_3, 123
Multiphoton processes, 16
Multiple reflection systems, 71, 72, 73, 77
Mutual exclusion rule, 108

$NaClO_3$, 136
Naphthalene, 129
Neutron diffraction, 106
$(NH_4)_2CuCl_4$, 109, 119, 121
Nielsen conditions, 72
Nitrogen trichloride, 157
Non-crossing rule, 95, 96
Normal coordinate, 85

Optic axis, 137–138
Optic(al) branch, 97, 112
Optical indicatrix, 134–136
Optic(al) modes, 94, 97, 103
Optically dense materials, 14
Orientation (of degeneracy), 117, 129
Oriented gas phase approximation, 84, 120, 126–131, 139
'Oscillations principales', 93
Oseen extinction theorem, 14
Overtone bands, 16, 17, 18, 126
Oxides, 106, 139

Paraffins, Raman of, 160, 161
$PdCl_4^{2-}$, 119, 138
$PdCl_6^{2-}$, 119, 139
Phase, 87, 92, 93
Phase-sensitive amplification, 34
Phonons, 93, 97
Photoelectric Raman spectrometers, Cary 81 system, 22, 37
  other commercial systems, 37–49
Photomultipliers, general performance characteristics, 32, 33
Photon counting, 34
Photosensitive materials, Raman of, 157
Plane group, 91, 107, 123
Point group operations, 89
Point symmetry, 84, 90, 106
Polaritons, 100
Polarizability, 8, 9, 15
  bond, 81
  tensor, 18
Polarization, of light, 4–6, 134–136
  of light, plane, 5
  of Raman lines, 61, 65–67, 69, 131–133, 160
Polarization scrambler, 67
Polarizing microscope, 134–136
Polymers, 85, 91, 100–101, 139
  Raman spectra, 166
  see also details under Raman spectra

Polypeptides, 92
Powders, 78–79, 153 *et seq*
Preresonance, 14
Pre-slit optical systems, advantages and disadvantages of 90° and 180° systems, 57
  90° arrangements, 28, 143
  Cary 81, 28
  Spex system, 48
Primitive cell, 88–89, 90, 94, 102, 109
Primitive translation, 86, 88, 90, 91, 92, 93, 97, 102, 105, 108, 109
  space, 93
Product rule, 100
Proper rotation, 111
Proteins, 92
$PtCl_4^{2-}$, 119, 138, 155
$PtCl_6^{2-}$, 119, 139

Qualitative analysis, 165
Qualitative Raman spectra, 56
Quantum theoretical treatment, 79–80
Quarter-wave plate, 8
Quasi-classical treatment, 11–12, 79
Q-spoiled Nd^{3+} laser, 194
Q-spoiled ruby laser, 205

Raman absorption, 16, 195
Raman intensity, 2, 61, 67–70, 79, 131–133
  temperature dependence, 12
Raman spectra (details given), aluminium tribromide, 151
  methylacetylene (gas), 146
  nitrogen trichloride, 157
  Nylons, 179
  polyethylene, 170
  polyethylene terephthalate, 179
  polyoxymethylene, 176
  polypropylene, isotactic, 172
    syndiotactic, 173
  polymethyl methacrylates, 178
  polytetrafluoroethylene, 180
  polyvinyl fluoride, 182
  polyvinylidene fluoride, 182
  sulphur monochloride, 54, 55
  xenon hexafluoride, 148
Raman spectra of polymers, experimental methods, 168
Raman tensor, 11, 20, 84, 126–133, 138–139
Rayleigh scatter, 8–9, 14
Recorders, 32, 53
  advantages of coupling, 53–55
Reduction formula, 112
Resolution of spectrometers, 49, 52
  limitation by sources, 143
Resonance denominator, 13
Resonance Raman effect, 1, 2, 13
Retardation plates, 6–8
Rotation, of axes, 127–128, 138–139, 208–209
  of molecules, 66, 115, 207–209
Rotation–inversion axes, 90
Rotation Raman spectra, 146
Rotation–vibration Raman spectra, 146
R-space, 87

Sample cells, 28–30
  Cary 81, 39
  Coderg, 41
  in gas-phase systems, 144, 145, 150
  for high-temperature gas-phase systems, 150
Sampling systems, for mercury arc sources, 22, 23
  for lasers, 26–30
  when sample is inside laser cavity, 35
  Spex system, 48
Schoenflies symbol, 90
Screw axes, 90, 91
Second-order (multiphonon) effects, 96, 103, 125–126
Selection rules, 17–20, 84, 103, 104, 112
Selenium monohalides, 159
Separable degeneracy, 112, 116
Separation of overlapping bonds by use of polarizer, 160
Set, 86
Short pulse generation, 194
Si–Ge, 105
Sign, of tensor component, 138–139
Silica gel, adsorption experiments, 183–189
Single-crystal spectroscopy, 84–141
Single scattering, 8, 14

Site group, 86, 91, 106–108, 109, 116–119, 126, 207
Small specimens, Raman of, 158, 159
$S_2O_6^{2-}$, 130, 139
Source, radiofrequency type, 22
  Toronto arc, 22, 143
  laser sources, 24–26
  comparison of lasers and discharge lamps, 26
Space group, 88, 89–91, 93, 107, 108, 109
Space phase difference, 87, 92, 93, 95
Specific heat, 93, 94, 96
Spectrographic detection, 22, 35
Spin waves, 140
SrF_2, 105
SrF_2–BaF_2, 105
Stimulated emission, 4, 15, 191–194
Stimulated Raman effect, 15, 191–194
Stoicheff absorption, 16, 195
Stokes, 2, 12, 17, 79
Stray light, discrimination in monochromators, 31, 53
Structure determination, 106, 155
Sub-lattice, 99, 105
Sub-structure, 102–104
Sulphur, rhombic, 129
Sulphur monohalides, 159
Supercells, 96, 102–104
Super-structure, 102–104
Symmetrically equivalent sets, 86, 108–109
Symmetry coordinates, 85, 87, 117, 123, 125
Symmetry operations, 127
  point, 89, 118–119
Symmetry type (species), 19, 85, 86, 87, 101, 103, 112, 116–119

$Tl_2Cl_9^{3-}$, 119
Toronto arc, construction and performance, 22, 143
Transformation of axes, 127–128, 138–139, 208–209
Transition moment, 1, 2, 10, 11
Translational modes, 207–209
Translational symmetry elements, 85, 108
Transverse mode, 97–100
Tysonite fluorides, 104, 106

U-centres, 105
Unit cells, 88–89, 101
  centred, 88, 109

Vibrations and symmetry, 85–87
Vibronic coupling, 80
Virtual state (level), 10, 13

Wave vector, 93, 97, 98, 100–101, 103, 125
Wavelength, of crystal mode, 93
Wavenumber or wavelength, calibration, 50
  drive system, 38
WOF_4, 106
Wolkenstein theory, 81–83, 130

X-ray methods, 104, 106

$ZnCl_4^{2-}$, 120, 130
ZnS, 99
Zones, Brillouin, 93, 94, 105, 112, 125